Materials Engineering

Materials Engineering

Richard Hammond

STATES
ACADEMIC PRESS
www.statesacademicpress.com

Materials Engineering
Richard Hammond
ISBN: 978-1-63989-344-7 (Hardback)

STATES
ACADEMIC PRESS

Published by States Academic Press,
109 South 5th Street,
Brooklyn, NY 11249, USA

Cataloging-in-Publication Data

Materials engineering / Richard Hammond.
p. cm.
Includes bibliographical references and index.
ISBN 978-1-63989-344-7
1. Materials. 2. Engineering. I. Hammond, Richard.
TA403 .M38 2022
620.11--dc23

Table of Contents

Preface

Materials engineering is an interdisciplinary branch of engineering which studies the various properties of different materials. It makes use of the principles of mechanical, electrical and civil engineering along with basic sciences to study the fundamental physical as well as chemical properties of the existing materials and then create new materials with desired properties. Some of the key areas of focus of this subject are atomic structure, crystallography, macrostructure and nanostructure. Emerging fields such as polymer and ceramic engineering are also its sub-branches. Polymer engineering deals with the study of materials such as thermoplastics, elastomers and thermosets, while ceramic engineering is concerned with the creation of objects from inorganic and non-metallic compounds. In the last few decades, material science, engineering has developed at a rapid pace. Some of the important innovations in the field of materials engineering are transistors, rockets, Kevlar, etc. Different approaches, evaluations and methodologies and advanced studies on materials engineering have been included in this book. The topics included in this book on materials engineering are of utmost significance and bound to provide incredible insights to readers. It will serve as a valuable source of reference for those interested in this field.

To facilitate a deeper understanding of the contents of this book a short introduction of every chapter is written below:

Chapter 1- Materials are the elements, substances or mixture of substances which something is composed of or can be made. Materials science is an interdisciplinary field which deals with the design and discovery of new materials. Crystallography, Tensile test, Polymer structures etc. are the fundamentals of materials science. This chapter introduces materials science and engineering to the reader in an easy-to-understand manner.

Chapter 2- There are various characterization techniques of materials like scanning electron microscopy, X-ray photoelectron spectroscopy, focused ion beam, X-ray diffraction, transmission electron microscopy and nanoindentation etc. This chapter has been carefully written to provide an easy understanding of the various characterization techniques of materials.

Chapter 3- All materials have specific and distinguishing properties. Physical properties are measurable like density, melting point, conductivity etc. Mechanical properties are the effects of different forces on the performance of a material. Various physical, electrical, biological, mechanical, thermal and chemical properties of materials are discussed in this chapter in detail.

Chapter 4- Chemical bond is a physical phenomenon in which the chemical substances are held together by attraction of atoms towards each other either by sharing or exchanging of electrons. There are different types of chemical bonds like covalent bond, ionic bond, coordinate covalent bond or dipolar bond etc. This chapter discusses in detail the different types of bonds and chemical bonding.

Chapter 5- Phase transformation takes place when one material changes its structure or composition. Phase diagram is used to represent the physical states of a substance under the influence of different conditions of temperature and pressure. All the diverse principles of phase transformation have been carefully analyzed in this chapter.

Chapter 6- The branch of materials science and engineering that is concerned with the extraction of metals from their ores and modification of metals for use is known as metallurgy. Extractive metallurgy involves the practice of extraction of metals from ores and refining them into a purer form while hydrometallurgy requires aqueous solutions to extract metals from ores. The topics elaborated in this chapter will help in gaining a better perspective about metallurgy.

I owe the completion of this book to the never-ending support of my family, who supported me throughout the project.

Richard Hammond

1

Materials Science: An Introduction

Materials are the elements, substances or mixture of substances which something is composed of or can be made. Materials science is an interdisciplinary field which deals with the design and discovery of new materials. Crystallography, Tensile test, Polymer structures etc. are the fundamentals of materials science. This chapter introduces materials science and engineering to the reader in an easy-to-understand manner.

Materials science is an interdisciplinary field concerned with the understanding and application of the properties of matter. Materials scientists study the connections between the underlying structure of a material, its properties, its processing methods and its performance in applications.

Materials science focuses on the relationship between the atomic and molecular structure of a material, the properties of the material (such as strength, electrical conductivity or optical properties), and ways in which the material is manufactured or processed into a shape or product. Another important part of materials science is how products can be improved using the knowledge of structure and processing techniques and how structure and processing influence material properties. This includes the study of the wide variety of classes of materials, from metals and composites to natural fibers and biological structures.

Materials science, engineering and technology include elements of applied physics and chemistry, along with aspects of electronics, manufacturing and production, combined with environmental, cost and quality considerations. The field is also an important part of forensic engineering and failure analysis with applications from aerospace to health. With significant media attention focused in recent years on new areas such as nanotechnology, biomaterials and electronic and optical systems, materials science has been propelled to the forefront of the public's consciousness. This leads materials technology, engineering and science to a new threshold of importance in the future of science and technology.

Materials are probably more significant in our culture than we realize. Transportation, housing, clothing communication, reaction and food production and virtually every segment of our daily lives is influenced by materials. Materials have contributed to the advancement of a number of technologies, including medicine & health, information & communication, national security & space, transportation, structural materials, arts & literature, textiles, personal hygiene, agriculture & food science & the environment. This inter-disciplinary interaction between the Material sciences and other fields in the development of new materials and their applications is to be understood well.

As the contribution of materials science and engineering to other disciplines increases, it will become necessary for scientists of all backgrounds to better understand it. Although it is not feasible for scientists to master a vast body of scientific knowledge over many disciplines, scientists must gain the skills that will allow them to master some specific topics. Materials Science and Engineering: (a) materials science, (b) materials engineering.

- Materials science involves investigating the relationships that exist between the structures and properties of materials.

- Materials engineering is based on the application of this structure-property correlations, in designing or engineering the structure of a material to produce a pre-determined set of properties.

From a functional perspective, the role of a materials scientist is to develop or synthesize new materials, whereas a materials engineer is called upon to create new products or systems using existing materials and/or to develop techniques for processing materials.

Elements of Materials Science and Engineering

There are four essential elements in materials science and engineering:

- Processing/synthesis
- Structure/composition
- Properties
- Performance/application

These four elements of Materials Science and Engineering is primarily concerned with the study of the basic knowledge of materials: the relationships between the composition/structure, properties and processing of materials. Materials engineering is mainly concerned with the use of this fundamental knowledge to design and to produce materials with properties that will meet the requirements of society. As subjects of study, materials science and materials engineering are very often closely related. The subject –materials science and engineering" combines both a basic knowledge and application and forms a bridge between the basic sciences (physics, chemistry and mathematics) and the various engineering disciplines, including electrical, mechanical, chemical, and civil and aerospace engineering.

The structure of a material usually relates to the arrangement of its internal components. Subatomic structure involves electrons within the individual atoms and interactions with their nuclei. On an atomic level, structure encompasses the organization of atoms or molecules relative to one another. The next larger structural realm, which contains large groups of atoms that are normally agglomerated together, is termed microscopic, meaning that which is subject to direct observation using some type of microscope. Finally, structural elements that may be viewed with the naked eye are termed macroscopic. Material structure can be classified as: macrostructure, microstructure, substructure, crystal structure, electronic structure and nuclear structure.

- Macro structure: The macrostructure of a material is examined by low-power magnification or naked eye. It deals with the shape, size and atomic arrangement in a crystalline material. In case of some crystals, e.g., quartz, external form of the crystal may reflect the internal symmetry of atoms. Macrostructure may be observed directly on a fracture surface or on a forging specimen. The individual crystals of a crystalline material can be visible, e.g. in a brass doorknob by the constant polishing and etching action of a human hand and sweat. Macrostructure can reveal flaws, segregations; cracks etc. by using proper techniques and one can save much expenses by rejecting defective materials at an early stage.
- Micro structure: This generally refers to the structure of the material observed under

optical microscope. Optical microscopes can magnify a structure about 1500 to 3000 times linear, without loss of resolution of details of the material structure. We may note that optical microscopes can resolve two lines separately when their difference of separation is 10–7 m (= 0.1m). Cracks, porosity, non-metallic inclusions within materials can be revealed by examining them under powerful optical microscope.

- Sub structure: When crystal imperfections such as dislocation in a structure are to be examined, a special microscope having higher magnification and resolution than the optical microscope is used. Electron microscope with magnifications 105 are used for this purpose. Another important modern microscope is field ion microscope, which can produce images of individual atoms as well as defects in atomic arrangements.
- Crystal structure: This reveals the atomic arrangement within a crystal. X-ray diffraction techniques and electron diffraction method are commonly used for studying crystal structure. It is usually sufficient to study the arrangement of atoms within a unit cell. The crystal is formed by a very large number of unit cells forming regularly repeating patterns in space.
- Electronic structure: This refers to the electrons in the outermost shells of individual atoms that form the solid. Spectroscopic techniques are commonly used for determining the electronic structure.
- Nuclear structure: This is studied by nuclear spectroscopic techniques, e.g., nuclear magnetic resonance (NMR).

Properties of Materials

A property is a material trait in terms of the kind and magnitude of response to a specific imposed stimulus. Generally, definitions of properties are made independent of material shape and size. The properties of engineering materials can be classified into two main groups:

- Physical
- Chemical

Virtually all important properties of solid materials may be grouped into six different categories:

1. Mechanical properties relate deformation to an applied load or force; examples include: elastic modulus or Young's modulus and strength; tensile and shear strengths, hardness, toughness, ductility, deformation and fracture behaviours, fatigue and creep strengths, wear resistance, etc. The important mechanical properties affecting the selection of a material are:

- Tensile Strength: This enables the material to resist the application of a tensile force. To withstand the tensile force, the internal structure of the material provides the internal resistance.
- Hardness: It is the degree of resistance to indentation or scratching, abrasion and wear. Alloying techniques and heat treatment help to achieve the same.
- Ductility: This is the property of a metal by virtue of which it can be drawn into wires or elongated before rupture takes place. It depends upon the grain size of the metal crystals.

- Impact Strength: It is the energy required per unit cross-sectional area to fracture a specimen, i.e., it is a measure of the response of a material to shock loading.
- Wear Resistance: The ability of a material to resist friction wear under particular conditions, i.e. to maintain its physical dimensions when in sliding or rolling contact with a second member.
- Corrosion Resistance: Those metals and alloys which can withstand the corrosive action of a medium, i.e. corrosion processes proceed in them at a relatively low rate are termed corrosion-resistant. g) Density: This is an important factor of a material where weight and thus, the mass is critical, i.e. aircraft components.

2. Thermal properties of solids can be represented in terms of heat capacity and thermal conductivity; the characteristics of a material, which are functions of the temperature, are termed its thermal properties. One can predict the performance of machine components during normal operation, if he has the knowledge of thermal properties. Specific heat, latent heat, thermal conductivity, thermal expansion, thermal stresses, thermal fatigue, etc., are few important thermal properties of materials. These properties play a vital role in selection of material for engineering applications, e.g. when materials are considered for high temperature service. Now, we briefly discuss few of these properties:

- Specific Heat: It is the heat capacity of a unit mass of a homogeneous substance. For a homogeneous body, $c = C/M$, where C is the heat capacity and M is the mass of the body. One can also define it as the quantity of heat required to raise the temperature of a unit mass of the substance through 1°C. Its units are cal/g/°C.
- Thermal Conductivity (K): This represents the amount of heat conducted per unit time through a unit area perpendicular to the direction of heat conduction when the temperature gradient across the heat conducting element is one unit. Truly speaking the capability of the material to transmit heat through it is termed as the thermal conductivity. The higher the value of thermal conductivity, the greater is the rate at which heat will be transferred through a piece of given size. Copper and aluminum are good conductors of heat and therefore, extensively used whenever transfer of heat is desired. Bakelite is a poor conductor of heat and hence used as heat insulator. The heat flow through an area A which is perpendicular to the direction of flow is directly proportional to the area (A) and thermal gradient (dt/dx).
- Thermal Expansion: All solids expand on heating and contract on cooling. Thermal expansion may take place as linear, circumferential or cubical. A solid which expands equally in three mutually orthogonal directions is termed as thermally isotropic. The increase in any linear dimension of a solid, e.g. length, width, height on heating is termed as linear expansion. The coefficient of linear expansion is the increase in length per unit length per degree rise in temperature. The increase in volume of a solid on heating is called cubical expansion. The thermal expansion of solids has its origin in the lattice vibration and lattice vibrations increases with the rise in temperature. Obviously, the thermal conductivity (K) and electrical conductivity (σ) vary in the same fashion from one material to another.
- Thermal Resistance (RT): It is the resistance offered by the conductor when heat flow due

to temperature difference between two points of a conductor. It is given by where H _ rate of heat flow and θ_1 and θ_2 are temperatures at two points (°C).

- Thermal Diffusivity (h): It is given by a material having high heat requirement per unit volume possesses a low thermal diffusivity because, more heat must be added to or removed from the material for effecting a temperature change.
- Thermal Fatigue: This is the mechanical effect of repeated thermal stresses caused by repeated heating and cooling. The thermal stresses can be very large, involving considerable plastic flow. We can see that fatigue failures can occur after relatively few cycles. The effect of the high part of the temperature cycle on the strength of material plays an important factor in reducing its life under thermal fatigue.

3. Magnetic properties demonstrate the response of a material to the application of a magnetic field. Materials in which a state of magnetism can be induced are termed magnetic materials. There are five classes into which magnetic materials may be grouped:

- Diamagnetic
- Paramagnetic
- Ferromagnetic
- Antiferromagnetic
- Ferrimagnetic

Iron, Cobalt, Nickel and some of their alloys and compounds possess spontaneous magnetisation. Magnetic oxides like ferrites and garnets could be used at high frequencies. Due to their excellent magnetic properties along with their high electrical resistivity these materials today, find use in a variety of applications like magnetic recording tapes, inductors and transformers, memory elements, microwave devices, bubble domain devices, recording hard cores, etc. Hysteresis, permeability and coercive forces are some of the magnetic properties of magnetic substances which are to be considered for the manufacture of transformers and other electronic components.

4. Electrical Properties: Electrical conductivity, resistivity, dielectric strength, the stimulus is an electric field are few important electrical properties of a material. A material which offers little resistance to the passage of an electric current is said to be a good conductor of electricity. The electrical resistance of a material depends on its dimensions and is given by: Usually resistivity of a material is quoted in the literature. Unit of resistivity is Ohm-metre. On the basis of electrical resistivity materials are divided as:

- Conductors
- Semiconductors
- Insulators

In general metals are good conductors. Insulators have very high resistivity. Ceramic insulators are most common examples and are used on automobile spark plugs, Bakelite handles for electric iron, plastic coverings on cables in domestic wiring.

5. Optical properties: The optical properties of materials, e.g. refractive index, reflectivity and absorption coefficient etc. affect the light reflection and transmission the stimulus is electromagnetic or light radiation.

6. Chemical Properties: These properties includes atomic weight, molecular weight, atomic number, valency, chemical composition, acidity, alkalinity, etc. These properties govern the selection of materials particularly in Chemical plant. Deteriorative characteristics relate to the chemical reactivity of materials. In addition to structure and properties, two other important components are involved in the science and engineering of materials namely, processing performance. With regard to the relationships of these four components, the structure of a material will depend on how it is processed. Furthermore, a materials performance will be a function of its properties.

Classification of Materials in Engineering

The traditional method is to classify them according to their nature into metals, ceramics, polymers and composites. The factors which form the basis of various systems of classifications of materials in material science and engineering are:

- The chemical composition of the material.
- The mode of the occurrence of the material in the nature.
- The refining and the manufacturing process to which the material is subjected to prior to acquiring the required properties.
- The atomic and crystalline structure of material.
- The industrial and technical use of the material.

Common engineering materials that fall within the scope of material science and engineering may be classified into one of the following six groups:

- Metals (ferrous and non-ferrous) and alloys
- Ceramics
- Organic Polymers
- Composites including Wood materials
- Semi-conductors
- Biomaterials
- Advanced Materials

Metals and Alloys

They are inorganic materials composed of one or more metallic elements. They may also contain a small number of non-metallic elements. All the elements are broadly divided into metals and non-metals according to their properties. Metals are element substances which readily give up electrons to form metallic bonds and conduct electricity. Some of the important basic properties of metals are:

- Metals are usually good electrical and thermal conductors.
- At ordinary temperature metals are usually solid.
- To some extent metals are malleable and ductile.
- The freshly cut surfaces of metals are lustrous.
- When struck metal produce typical sound.
- Most of the metals form alloys when two or more pure metals are melted together to form a new metal whose properties are quite different from those of original metals, it is called an alloy.

Metals usually have a crystalline structure and are good thermal and electrical conductors. Many metals are strong and ductile at room temperature and maintain good strength at high and low temperatures. Metallic materials possess specific properties like plasticity and strength. Few favourable characteristics of metallic materials are high lustre, hardness, resistance to corrosion, good thermal and electrical conductivity, malleability, stiffness, the property of magnetism, etc. Metals may be magnetic, non-magnetic in nature.

These properties of metallic materials are due to:

- The atoms of which these metallic materials are composed.
- The way in which these atoms are arranged in the space lattice.

Metallic materials are typically classified according to their use in engineering as under:

- Pure Metals: They are obtained by refining the ore. Mostly, pure metals are not of any use to the engineers. However, by specialised and very expensive techniques, one can obtain pure metals (purity ~ 99.99 %), e.g. aluminum, copper, etc.
- Alloyed Metals: Alloys can be formed by blending two or more metals or at least one being metal. The properties of an alloy can be totally different from its constituent substances, e.g. 18-8 stainless steel, which contains 18 %, chromium and 8 % nickel, in low carbon steel, carbon is less than 0.15 % and this is extremely tough, exceedingly ductile and highly resistant to corrosion.
- Ferrous Metals: Iron is the principal constituent of these ferrous metals. Ferrous alloys contain significant amount of non-ferrous metals. Ferrous alloys are extremely important for engineering purposes. On the basis of the percentage of carbon and their alloying elements present, these can be classified into following groups:
    - Mild Steels: The percentage of carbon in these materials range from 0.15 % to 0.25 %. These are moderately strong and have good weldability. The production cost of these materials is also low.

    - Medium Carbon Steels: These contains carbon between 0.3 % to 0.6 %. The strength of these materials is high but their weldability is comparatively less.
    - High Carbon Steels: These contains carbon varying from 0.65 % to 1.5 %. These materials get hard and tough by heat treatment and their weldability is poor. The steel formed in which carbon content is up to 1.5 %, silica up to 0.5%, and manganese up to 1.5 % along with traces of other elements is called plain carbon steel.
    - Cast Irons: The carbon content in these substances vary between 2 % to 4%. The cost of production of these substances is quite low and these are used as ferrous casting alloys.
- Non-Ferrous Metals: Out of several non-ferrous metals only seven are available in sufficient quantity reasonably at low cost and used as common engineering metals. These are aluminum, tin, copper, nickel, zinc and magnesium. Some other non-ferrous metals, about fourteen in number, are produced in relatively small quantities but these are of vital importance in modern industry. These include chromium, mercury, cobalt, tungsten, vanadium, molybdenum, antimony, cadmium, zirconium, beryllium, niobium, titanium, tantalum and manganese.
- Sintered Metals: These materials possess very different properties and structures as compared to the metals from which these substances have been cast. Powder metallurgy technique is used to produce sintered metals. The metals to be sintered are first obtained in powered form and then mixed in right calculated proportions. After mixing properly, they are put in the die of desired shape and then processed with certain pressure. Finally, one gets them sintered in the furnace.
- Clad Metals: A sandwich of two materials is prepared in order to avail the advantage of the properties of both the materials. This technique is termed as cladding. Using this technique stainless steel is mostly embedded with a thick layer of mild steel, by rolling the two metals together while they are red hot. This technique will not allow corrosion of one surface. Another example of the use of this technique is cladding of duralumin with thin sheets of pure aluminum. The surface layers, i.e. outside layers of aluminum resist corrosion, whereas inner layer of duralumin imparts high strength. This technique is relatively cheap to manufacture.

Ceramics

Ceramics are inorganic materials consisting of both metallic and non-metallic elements bonded together chemically. Ceramics can be crystalline, non-crystalline or a mixture of both. Generally, they have high melting points and high chemical stabilities. They also have high hardness and high temperature strength but tend to be brittle. Ceramics are usually poor electrical conductors.

Polymers

Polymers are organic materials which consist of long molecular chains or networks containing carbon. Most polymers are non-crystalline, but some consist of mixtures of both crystalline and non-crystalline regions. They typically have low densities and are mechanically flexible. Their mechanical properties may vary considerably. Most polymers are poor electric conductors due to the nature of the atomic bonding.

Composites

Composites are mixtures of two or more types of materials. The constituent elements in a composite retain their identities (they do not dissolve or merge completely into each other) while acting in concert to provide a host of benefits such as light weight, high strength, corrosion resistant, high strength-to-weight ratio, directional strength - tailor mechanical properties, high impact strength, high electric strength (insulator), radar transparent, non-magnetic, low maintenance, long-term durability, parts consolidation, dimensional stability, small to large part geometry – styling/design – sculptural form, customized surface finish, rapid installation. Usually, they consist of a matrix phase and a reinforcing phase. They are designed to ensure a combination of the best properties of each of the component materials. There is also an increasing trend to classify engineering materials into two further categories: structural materials and functional materials. Structural materials, as the name indicates, are materials used to build structures, bodies and components. For instance, in a car the body, frame, wheels, seats, inside lining, engine and various mechanical transmission parts are all constructed from structural materials.

Semi-Conductors

These are the materials which have electrical properties that are intermediate between the electrical conductors and insulators. The electrical characteristics of semi-conductors are extremely sensitive to the presence of minute concentrations of impurity atoms; these concentrations may be controlled over very small spatial regions. Semi-conductors form the backbone of electronic industry. The semi-conductors have made possible the advent of integrated circuitrythat has totally revolutionized the electronics and computer industries. They affect all walks of life whether it is communications, computers, biomedical, power, aviation, defence, entertainment, etc. The field of semi-conductors is rapidly changing and expected to continue in the next decade. Organic semi-conductors are expected to play prominent role during this decade. Diamond as semiconductor will also be important. Optoelectronic devices will provide three dimensional integration of circuits, and optical computing.

The Importance of Engineering Materials in Our Present World

Development of new materials has followed a number of different pathways, depending on both the nature of the problem being pursued and the means of investigation. Breakthroughs in the discovery of new materials have ranged from pure serendipity, to trial-and-error approaches, to design by analogy to existing systems. These methodologies will remain important in the development of materials but as the challenges and requirements for new materials become more complex, the need to design and develop new materials from the molecular scale through the macroscopic final product will become increasingly important. The use of molecular modeling and the engineering of new materials into useable forms or devices are of particular importance.

Current Trends and Advances in Materials

Timber, steel and cement are the materials which are widely used for engineering applications in huge quantities. The consumption of steel in any country is considered as an indicator of its economic well-being. For high temperature applications. Newer materials for combined resistance to high temperature and corrosion are increasing rapidly and material scientists and engineers are busy in developing such materials. Different kinds of ceramics, though difficult to shape and machine, are finding demand for their use at high temperatures. Recently prepared new metallic materials in conjunction with new processing techniques as isostatic pressing and isothermal forging are capable of imparting better fatigue properties to aircraft components. Powder metallurgy technique while producing finished surfaces and cutting down metal cutting cost is much capable of imparting improved mechanical properties under different loading conditions. Surprisingly, rapid cooling technology achieving cooling rates in the vicinity of one million degree Celsius per second and this is being used to produce metal powders which can be used in such product producing techniques as powder metallurgy and hot isostatic pressing to obtain temperature resistant parts.

Nowadays, metallurgists have produced several molybdenum and aluminium alloys as well as alloys of titanium and nickel to meet anticorrosion properties at elevated temperatures. Polymeric materials are growing at annual rate of 9% and have grown in volume more than any other material. In several applications plastics have replaced metals, wood, glass and paper. A new trend in plastic technology is the production of synergistic plastic alloys which have better properties than individual members producing the alloy. Recent discovery of plastic conductors may have wider impact in near future. The major drawback of ceramics is the brittleness and difficulty in cutting and shaping. When mixed with metal powder like molybdenum, ceramic produce cements, which are expected to be useful cutting materials. Tool bits of cements are expected to find various applications in attaining high cutting speeds and producing better surface finish. Alumina, a well-known ceramic is expected to be successfully reinforced with fibres of molybdenum. Due to micro cracking of molybdenum fibres, the attempts to achieve better strength in such composite ceramics have not been successful yet. However, such composites have been found to exhibit better impact and thermal shock resistance.

Further Advances in Materials Development from View Point of New and Advanced Materials

Recent development especially from the point of view of new and advanced materials could be classified into:

Advanced Materials

The materials that are utilised in high-technology (or high-tech) applications are sometimes called advanced materials. By high technology we mean a device or product that operates or functions using relatively intricate and sophisticated principles; for example, electronic equipment (VCRs, CD players, etc.), computers, fiber optic systems, spacecraft, aircraft and military rocketry. These advanced materials are typically either traditional material whose properties have been enhanced or newly developed high performance materials. Furthermore, advanced materials may be of all material types (e.g., metals, ceramics, and polymers) and are normally relatively expensive. In subsequent chapters are discussed the properties and applications of a good number of advanced materials—for example, materials that are used for lasers, ICs, magnetic information storage, liquid crystal displays (LCDs), fiber optics, and the thermal projection system for the space shuttle orbiter.

Smart Materials

Smart or intelligent materials form a group of new materials now being developed that will have a significant influence on many of our technologies. In addition, the concept of smart materials is being extended to rather sophisticated systems that consist of both smart and traditional materials. The field of smart materials attempts to combine the sensor, actuator and the control circuit on as one integrated unit. Actuators may be called upon to change shape, position, natural frequency, or mechanical characteristics in response to changes in temperature, electric fields, and magnetic fields. The combined system of sensor, actuator and control circuit on as one IC unit, emulates a biological system. These are known as smart sensors, microsystem technology (MST) or micro-electro mechanical systems (MEMS). Materials/devices employed as sensors include optical fibers, piezoelectric materials, and MEMS. MEMS devices are small in size, light weight, low cost, reliable with large batch fabrication technology. They generally consist of sensors that gather environmental information such as pressure, temperature, acceleration etc., integrated electronics to process the data collected and actuators to influence and control the environment in the desired manner. The MEMS technology involves a large number of materials. Silicon forms the backbone of these systems also due to its excellent mechanical properties as well as mature micro-fabrication technology including lithography, etching, and bonding. Other materials having piezoelectric, piezo-resistive, ferroelectric and other properties are widely used for sensing and actuating functions in conjunction with silicon. The field of MEMS is expected to touch all aspects of our lives during this decade with revolution in aviation, pollution control, and industrial processes.

Nano-Structured Materials and Nanotechnology

Nanotechnology is a field that deals with control of structures and devices at atomic, molecular and supermolecular levels as well as the efficient use and manufacture of these devices. Key areas in Nanotechnology are:

- Nano-medicine for disease detection and treatment.
- Nano-engineered materials for improved agriculture.
- Nanotechnology for energy.
- Nano porous materials for water filtration.

Nanostructured materials are those materials whose structural elements clusters, crystallites or molecules have dimensions in the range of 1-100 nm. These small groups of atoms, in general, go by different names such as nanoparticles, nano-crystals, quantum dots and quantum boxes. One finds a remarkable variation in fundamental electrical, optical and magnetic properties that occur as one progresses from an infinitely extended solid to a particle of material consisting of a countable number of atoms. The various types of nanostructured materials which has been considered for applications in opto-electronic devices and quantum- optic devices are nano-sized powders of silicon, silicon-nitride (SiN), silicon-carbide (SiC) and their thin films. Some of these are also used as advanced ceramics with controlled micro structures because, their strength and toughness increase when the grain size diminishes. Carbon based nanomaterials and nanostructures including fullerenes and nanotube plays an increasingly significant role in nanoscale science and technology.

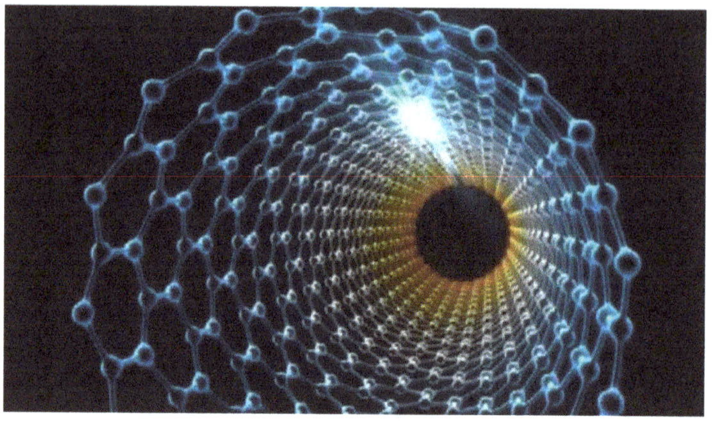

Nano materials structure.

Quantom Dots (QDS)

Rapid progress in the fabrication of semiconductor structures has resulted into the reduction of 3D bulk systems to 2D & 1D, and ultimately to 0D systems. Quantum dots represent the ultimate reduction in the dimensionality of semiconductor devices. These are 3D semiconductor structures only nanometer in size confining electrons and holes. QDs operate at the level of single electron which is certainly the ultimate limit for an electronic device and are used as the gain material in lasers. QDs are used in quantum dot lasers, QD memory devices, QD photo detectors and even quantum cryptography. The emission wavelength of a quantum dot is a function of its size. So by making dots of different sizes, one can create light of different colors.

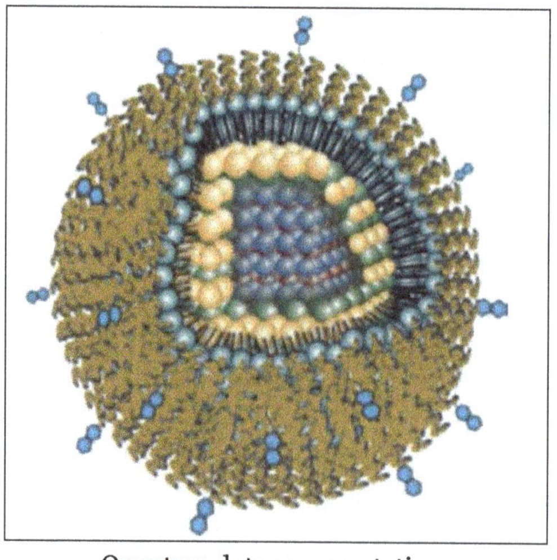

Quantum dots representation.

Spintronics

A revolutionary new class of semiconductor electronics based on the spin degree of freedom could be created. The study of electron spin in materials is called spintronics. Spintronics is based on the direction of spin- and spin coupling. Every appliance ranging from electric bulb to laptop computer works on the principle of transport of electric charge carriers-electrons. The electrons have both charge and spin. The spin of the electrons could greatly enhance the particles usefulness. Presently, the semiconductor technology is based on the number of charges and their energy. The electron can be assumed as tiny rotating bar magnet with two possible orientations: spin-up or spin-down. An applied magnetic field can flip electrons from one state to another. Obviously, spin can be measured and manipulated to represent the 0's and 1's of digital programming analogous to the –current on and current off states in a conventional silicon chip. The performance of conventional devices is limited in speed and dissipation, whereas, spintronic devices are capable of much higher speed at very low power. Spintronic transistors may work at a faster speed, are also smaller in size and will consume less power. The electron spin may exist not only in the up or down state but also in infinitely many intermediate states because of its quantum nature depending on the energy of the system. This property may lead to highly parallel computation which could make a quantum computer work much faster for certain types of calculations. The mixed state could form the base of a computer, built around not only on binary bits but the quantum bits or cubit. It is any combination of a 1 or a 0. Scientists are now trying to use the property of the electron-like spin rather than charge to develop new generation of microelectronic devices which may be more versatile and robust than silicon chips and circuit elements. Spins appear to be remarkably robust and move effectively easily between semi-conductors.

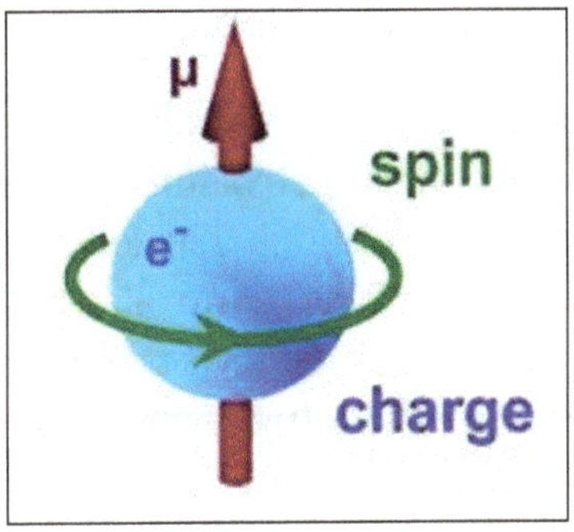

Spintronics

Fermionic Condensate Matter

Very recently scientists had created a new form of matter called a fermionic condensate matter and predicted it could lead to the next generation of superconductors. Solids, liquids, gases, plasma and Bose-Einstein condensate, fermionic condensate matter is a scientific breakthrough in providing a new type of quantum mechanical behaviour. It is related to Bose-Einstein condensate. However, new state is not a superconductor but it is really something in between these two states.

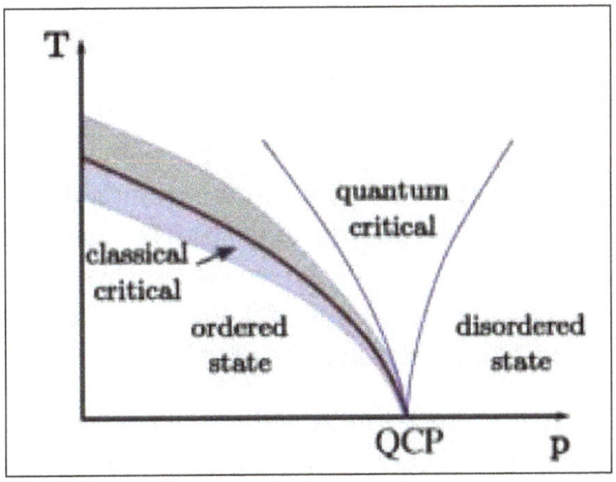

Fermionic matter.

Breakthroughs in Materials Development

The Material Science has made great strides over the past several decades in the development of novel materials. Although the following is not meant to be an exhaustive list of such breakthroughs, these examples point to the range of materials and their applications.

Examples such as Teflon serve to show how the chemical sciences have contributed indispensable materials to everyday use. More recently, the development of thermoplastics and structural polymers has had an increasing influence on applications ranging from construction to national defense. New paints and coatings, clothing fibers, and photographic films have all benefited from the development of new materials. There are newer polymeric materials whose commercial impact has yet to be realized. Work on semi-conductive and conductive polymers have made great strides, but further work is necessary. Synthesis of amphiphilic dendritic block copolymers that are designed to form ultrathin organic films have also had major advances, other promising materials, from polymers for drug delivery to tissue engineering, have the potential to benefit the biomedical field but are still in a relatively early stage of development.

1. Catalysis Advances: Catalysis Advances in new materials cover a wide range of applications. Zeolites and pillared clays have had a huge impact on the petroleum industry. New zeolites with specified properties continue to be developed with various utilities. Ziegler-Natta catalysts allow the preparation of billions of pounds per year of organic polymers with controlled molecular structures and useful material properties. This method is also useful because it allows the synthesis of polymers that cannot be produced in a practical manner by any other method. In the energy and transportation sector, catalysis has been an especially fruitful area of research. As a result, supported gold catalysts have been developed. In addition, selective oxidation of carbon monoxide has been achieved and a gold transition- metal oxide has been developed that provides very active NOx reduction as well as hydrocarbon oxidation. The platinum particles serves to complete the oxidation of hydrocarbons and carbon monoxide to carbon dioxide, while rhodium converts nitrogen oxides to nitrogen and oxygen. The use of supramolecular organic templates containing appropriate surface functionalities to regulate the nucleation and growth of inorganic magnets, semiconductors, and catalysts is significant as well.

2. Electronics: This broad category has benefited from many breakthroughs in the development of new materials. Perhaps no recent advance has had a greater impact in this area than the creation of chemically amplified photo resist. Photo resist, resins containing photo chemically active polymers, can be coated on a wafer and irradiated using photons (photolithography), electrons (electron beam lithography), or X-rays (X-ray lithography). These developments have had considerable impact on computer chip production. In the field of telecommunications, high-temperature superconductors and ceramic materials containing copper-oxide planes have potential uses in communications shielding. The development of new instrumentation is essential both in characterizing materials and in exploring their potential applications. Matrix-assisted laser desorption ionization time-of-flight mass spectrometry (MALDI-TOF), a mass spectrometry technique that uses laser light to ablate un-fragmented polymer molecules mixed with an organic acid matrix into a time of- flight mass spectrometer, is finding increasing application in the polymer community.

3. Composites: Composites have recently found wide application in various area of life such as Aircraft, Commercial, pleasure and military aircrafts, including components for aerospace and related applications; Appliance, Business. Composite applications for the household and office

including appliances, power tools, business equipment, etc; Automotive/Transportation: The largest of the markets, products include parts for automobiles, trucks, rail and farm applications; Civil Infrastructure: A relatively new market for composites, these applications include the repair and replacement of civil infrastructure including buildings, roads, bridges, piling, etc. Construction: Includes materials for the building of homes, offices, and architectural components. Products include swimming pools, bathroom fixtures, wall panels, roofing, architectural cladding; Consumer products: Sports and recreational equipment such as golf clubs, tennis rackets, snowmobiles, mobile campers, furniture, microwave cookware; Corrosion-Resistant and Chemical-Resistant Equipment: Tanks, ducts and hoods, pumps, fans, grating, chemical processing, pulp and paper, oil and gas and water/wastewater treatment markets; Electrical: Components for both electrical and electronic applications such as pole line hardware, substation equipment, microwave antennas, printed wiring boards, etc; Marine: Products for commercial, pleasure and naval boats and ships.

Modern Materials: Needs and Challenges

In spite of the tremendous progress made in the field of material science within the past few years, there still remains technological challenges including the development of more sophisticated and specialized materials, as well as the impact of materials production on ecosystem. Nuclear (fission as well as fusion) energy holds some promise, but the solutions to the many key problems that remain will, necessarily, involve materials from fuels to containment structures to facilities for the disposal of radioactive waste. Progress in fusion research has been incredibly rapid in recent years. There has been major progress in fusion materials and technology, with prototypes of the key components of a fusion power plant built and successfully tested. Obviously, new high strength, low density structural materials remain to be developed, as well as materials that have higher temperature capabilities, for use in engine components. Furthermore, there is an urgent need to find new, economical sources of energy and to make use of present energy resources more economically. Hydrogen seems to be the fuel of the future. Hydrogen offers the greatest potential environment and energy supply benefits. Like electricity, hydrogen is a versatile energy carrier that can be made from a variety of widely available primary energy sources including natural gas, coal, biomass, wastes, sunlight, wind, and nuclear power. Although hydrogen production techniques do exist, further optimization is desirable for use in energy systems with zero carbon emissions. Materials will undoubtedly play a significant role in these developments, We know that environment quality depends on our ability to control air and water pollution. Pollution control techniques employ various materials. There is a need to improve material processing and refinement methods so that they produce less environmental degradation. Toxic substances are produced during manufacturing processes of some materials and therefore we have to consider the ecological impact of their disposal. There are many materials which we use are derived from resources that are non-renewable. These include polymers for which the prime raw material is oil, and some metals. These non-renewable resources are gradually becoming depleted, which necessitates:

- The search of additional reserves.
- The development of new materials having comparable properties with less adverse environmental impact.
- Increased recycling efforts and the development of new recycling technologies.

Materials

A material is defined as a substance (most often a solid, but other condensed phases can be included) that is intended to be used for certain applications. There are a myriad of materials around us – they can be found in anything from buildings to spacecraft. On the basis of chemistry and atomic structure, materials are classified into three general categories:

- Metals (metallic elements).
- Ceramics (compounds between metallic and non-metallic elements).
- Polymers (compounds composed of carbon, hydrogen, and other non-metallic elements).

12 grams polycrystalline germanium.

In addition, composites are composed of at least two different material types. New and advanced materials that are being developed include nanomaterials, biomaterials, and energy materials to name a few.

Materials for Nuclear Engineering

An understanding of material science is essential for power plant personnel to understand why a material was selected for certain applications within their facility. Almost all processes that take place in the nuclear facilities involve the use of specialized metals. A basic understanding of material science is necessary for nuclear facility operators, maintenance personnel, and the technical staff to safely operate and maintain the facility and facility support systems. Our goal here will be to introduce material engineering of nuclear reactors. The knowledge of thermophysical and nuclear properties of materials is essential for designing nuclear power plants.

1. Metals: Metal is a material (usually solid) comprising one or more metallic elements (e.g., iron, aluminum, copper, chromium, titanium, gold, nickel), and often also non-metallic elements (e.g., carbon, nitrogen, oxygen) in relatively small amounts. The unique feature of metals as far as their structure is concerned is the presence of charge carriers, specifically electrons. This feature is given by the nature of metallic bond. In metallic bond, the atoms do not share or exchange electrons to bond together. Instead, many electrons (roughly one for each atom) are more or less free to move

throughout the metal, so that each electron can interact with many of the fixed atoms. The electrical and thermal conductivities of metals originate from the fact that their outer electrons are delocalized.

2. Ceramics: A ceramic is a solid material comprising an inorganic compound of metal, non-metal or metalloid atoms primarily held in ionic and covalent bonds. Common examples are earthenware, porcelain, and brick. In nuclear industry, uranium dioxide is a ceramic refractory uranium compound, in many cases used as a nuclear fuel.

3. Polymers: Polymers are compounds (macromolecules) composed of carbon, hydrogen, and other non-metallic elements. Polymers range from familiar synthetic plastics such as polystyrene to natural biopolymers such as DNA and proteins that are fundamental to biological structure and function. Some common and familiar polymers are polyethylene (PE), nylon, polycarbonate (PC), polystyrene (PS), and silicone rubber.

In addition, composites are composed of at least two different material types. Another materials category is the advanced materials that are used in high-tech applications, including:

- Semiconductors: In general, semiconductors are materials, inorganic or organic, which have the ability to control their conduction depending on chemical structure, temperature, illumination, and presence of dopants. The name semiconductor comes from the fact that these materials have an electrical conductivity between that of a metal, like copper, gold, etc. and an insulator, such as glass.
- Biomaterials: In general, a biomaterial is any substance that has been engineered to interact with biological systems for a medical purpose. These materials must not produce toxic substances and must be compatible with body tissues (i.e., must not cause adverse biological reactions).
- Smart materials: Smart materials are materials that sense and respond to changes in their environments in predetermined manners.
- Nanomaterials: Nanomaterials are materials that have structural features on the order of a nano-meter, some of which may be designed on the atomic/molecular level.

Biomaterials

Biomaterials are materials (synthetic and natural; solid and sometimes liquid) that are used in medical devices or in contact with biological systems. Biomaterials as a field has seen steady growth over its approximately half century of existence and uses ideas from medicine, biology, chemistry, materials science and engineering. There is also a powerful human side to biomaterials that considers ethics, law and the health care delivery system.

Although biomaterials are primarily used for medical applications, they are also used to grow cells in culture, to assay for blood proteins in the clinical laboratory, in processing biomolecules in biotechnology, for fertility regulation implants in cattle, in diagnostic gene arrays, in the aquaculture of oysters and for investigational cell-silicon "biochips." The commonality of these applications is the interaction between biological systems and synthetic or modified natural materials.

Biomaterials are rarely used on their own but are more commonly integrated into devices or implants. Thus, the subject cannot be explored without also considering biomedical devices and the biological response to them. Some common medical devices comprised of biomaterials are illustrated here:

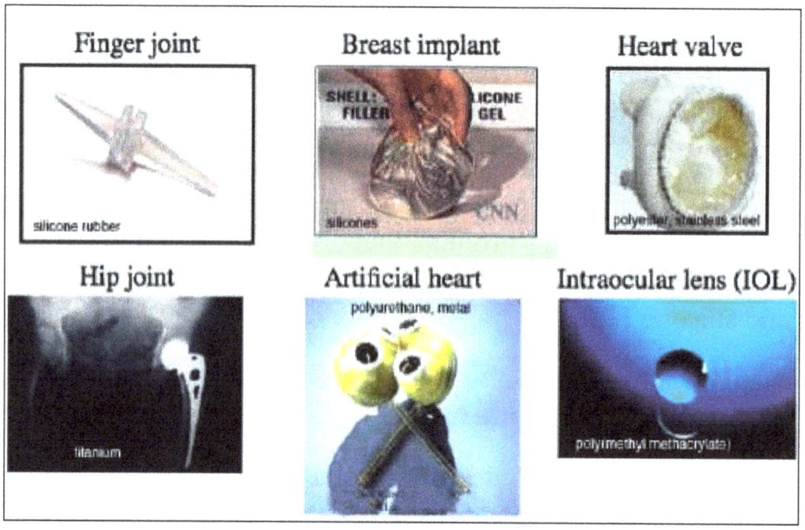

Biomaterials can be metals, ceramics, polymers, glasses, carbons, and composite materials. Such materials are used as molded or machined parts, coatings, fibers, films, foams and fabrics. A biomaterial is a nonviable material used in a medical device, intended to interact with biological systems.

If the word "medical" is removed, this definition becomes more general and still quite useful. If the word "nonviable" is removed, the definition becomes even more general and can address new tissue engineering and hydrid artificial organ applications where living cells are used.

A complementary definition needed to understand an important aspect of biomaterials is that of "biocompatibility." Biocompatibility is the ability of a material to perform with an appropriate host response in a specific application.

"Appropriate host responses" include lack of blood clotting, resistance to bacterial colonization and normal healing. Examples of specific applications include a hemodialysis (artificial kidney) membrane, a urinary catheter or a hip joint prosthesis. It is interesting that the hemodialysis membrane might be in contact with the patient's blood for 3 hours, the catheter may be inserted for a week and the hip joint may be in place for the life of the patient. The definitions introduce us to considerations that set biomaterials apart from most materials considered in materials science.

Characteristics of Biomaterials as a Field

It's Multidisciplinary

Biomaterials science brings together researchers from diverse academic backgrounds. They must communicate clearly. Some disciplines that intersect in the development, study and application of biomaterials include: bioengineer, chemist, chemical engineer, electrical engineer, mechanical engineer, materials scientist, biologist, microbiologist, physician, veterinarian, ethicist, nurse, lawyer, regulatory specialist and venture capitalist. This is only a partial list.

It Uses Many Diverse Materials

Biomaterials researchers and clinicians using biomaterials will have an appreciation of materials science and chemistry. Many different synthetic and modified natural materials are used in biomaterials and some understanding of the differing properties of these materials is important. A heart valve may be fabricated from polymers, metals, and carbons. A hip joint might be fabricated from metals and polymers (and sometimes ceramics) and will be interfaced to the body via a polymeric bone cement. In these examples, a single device uses many different materials, each with special properties and biological interactions.

The End Product is the Development of Devices

Biomaterials by themselves do not make a useful clinical therapy. The materials have to be fabricated into devices. This is typically an engineer's role, but the engineer might work closely with synthetic chemists to optimize materials properties and physicians to ensure that the device is useful in clinical applications.

The Magnitude of the Field is Generally Unappreciated

The magnitude and economic scope of the contemporary organ replacement enterprise are much larger than is generally recognized. In the year 2000, the lives of over 20 million patients will be sustained, supported or significantly improved by functional organ replacement. The impacted population grows at over 10% per year. Worldwide, first-year and follow-up costs of organ replacement and prostheses exceeds $300 billion US dollars per year and represents between 7% and 8% of total worldwide healthcare spending. In the United States, the costs of therapies enabled by organ replacement technology exceed 1% of the gross national product. The costs are also impressive when considering the needs of the individual patient. Thus, the cost of a substitute heart valve is roughly $4000. The surgery to implant the device costs approximately $60,000. Reoperation for replacing a failed valve will have these same costs and reoperations occur for 10% of all valve replacements.

Success and Failure are seen with Biomaterials and Medical Devices

Most biomaterials and medical devices perform satisfactorily, improving the quality of life for the recipient or saving lives. Still, manmade constructs are never perfect. Manufactured devices have a failure rate. Also, all humans differ in genetics, gender, body chemistries, living environment and physical activity. Furthermore, physicians also differ in their "talent" for implanting devices. The other side to the medical device success story is that there are problems, compromises and complications that occur with medical devices. Central issues for the biomaterials scientist, manufacturer, patient, physician and attorney are: (1) what represents good design; (2) who should be responsible when devices perform "with an inappropriate host response;" and (3) what are the cost/risk or cost/benefit ratios for the implant or therapy?

An example involving left ventricular assist devices (LVADs, sometimes incorrectly called artificial hearts) helps to clarify these issues. There are many complications with LVADs including clotting, strokes, blood damage and bacterial infection. A clinical trial called Randomized Evaluation of Mechanical Assistance for the Treatment of Congestive Heart Failure (REMATCH) led to following important statistics. Patients with an implanted Heartmate® LVAD (Thoratec Laboratories) had a 52% chance of surviving for one year, compared with a 25% survival rate for patients who took medication (the common therapy for congestive heart failure). Survival for two years in patients with the Heartmate® was 23% versus 8% in the medication group. Also, the LVAD enhanced the quality of life for the patients – they felt better, were less depressed, and were mobile. Note that patients participating in the REMATCH trial were not eligible for a heart transplant. In the case of the LVAD, long-term clinical complications associated with imperfect performance of biomaterials does not preclude clinical success overall. The LVAD is better than the next best therapy.

These five characteristics of biomaterials science–multidisciplinary, multi-material, needdriven, substantial market and risk-benefit–color the field of biomaterials.

Metallic Glasses

In our day-to-day life, we use several types of products made from different materials. Based on the area of application, the desired properties of equipment for their production and processing can vary. According to this, material selection takes place in the way to supply the best possible outcomes compared to others. In recent decades, aluminum, steel and plastics have been the most commonly used materials. The aluminum is considered as the best choice for automobile and structural object for its low density and high specific strength, whereas, due to good strength and cost-effectiveness, steel is the most preferred material for structural applications like construction, railway industry etc. Likewise, plastics are used for various home and kitchen appliances and also for the interior design of buildings, vehicles etc. Plastics are very adaptable materials because of their easy processing, but characterized by low strength compared to metals. On the other hand, aluminum and steel lose the battle with plastics in area of their processing in order to produce very intricate shapes. Because of the high processing temperature, metal products may have more defects over plastics. In this sense, metallic glasses compete with both metals and plastics. These materials have good strength and toughness compared to plastics and can be formed in any desired intricate design compared to metals. Also they possess high corrosion and wear resistance. So, we can say that metallic glasses are the materials having the good properties of metals (like steel and aluminum) as well as good adaptability like plastics. In metallic glasses, during deformation, dislocation movements occurs, shearing a localization of atoms, but in the case of crystalline materials several defects occur weakening them.

As promising materials for different applications, metallic glasses are preferred over metals, ceramics, magnetic and some other types of existing materials due to their enhanced properties. Some of the important reasons for which we consider these glasses for specific applications are discussed in the followings:

- Metallic glasses have no long-range of ordering like crystalline materials. It develops more homogeneity inside the material because defects like point defects, dislocations and stacking faults are absent.
- These materials possess very high strength in the elastic region. It can be declared as a good yielding strength of the material which is higher than steel.
- Because of the good homogeneity of atoms in metallic glasses, very good corrosion resistance is achieved along with good wear resistance.

- Ordinary silica glasses are brittle in nature unlike the metallic glasses which are very tough materials.
- These materials have good luster and mirror effects but they are opaque.
- The metallic glasses are very hard materials and their fracture resistance is much better compared to ceramics.
- Because of the metallic atoms, these glasses possess significant magnetic effects. It helps to easily magnetize or demagnetize these materials. Metallic glasses with soft magnetism have very small hysteresis loop. Due to the narrow hysteresis, in these glasses hysteresis loss is minimized.
- Because of the amorphous structure of metallic glasses, their electrical resistivity is higher resulting in less eddy current loss during its application.

The good adaptation of these materials for different industrial applications is a consequence of combinations of those properties. The discovery of metallic glass in 1960 motivated scientists to research and manufacture this kind of materials. The researchers got the non-crystalline structure in Au-Si alloys. Rapid quenching of those alloys from their liquid state was conducted by the gun technology. They formed a very thin layer of metallic glass over a cold copper substrate. The reason to take copper as the substrate is because of its good thermal conductivity. After that, people are continuously discovering various metallic glasses with different compositions of elements. After 2000 AD, people are making varieties of metallic glasses and its demand is increasing for industrial applications. In the 1990s, the development of different BMGs (Bulk Metallic Glasses) based on late transition metals (LTM) started. A. Inoue et al. successfully developed the Fe–Al–Ga–P–C–B BMGs in 1995. Today the availability and cost effectiveness are the two major factors in selecting and production of such materials. For the stabilization of a supercooled metallic liquid the three rules were proposed:

- Multicomponent BMG systems must consist of at least three or more elements.
- Atomic size of the constituent elements should have a significant size difference (greater than 12%).
- Heat of mixing of the elements should be negative.

General Description of Metallic Glasses

According to atomic arrangement, we can categorize the existing and man-made solid materials into two main groups: crystalline and amorphous. When there is a proper ordered arrangement of atoms then we say it is a crystalline material. If there is a random arrangement of atoms, then the material is called amorphous. The atomic arrangements of crystalline and amorphous materials are shown in Figure. To get such randomness, the sizes of the atoms are very important. Much difference in the atomic radius of the components leads to more randomness in the atomic arrangement. Glass forming is majorly concerned with the study of crystallization of materials in order to avoid crystallization. When metallic alloys are cooled at a very fast rate, possibilities of getting an ordered arrangement are poor.

The glass transition temperature (Generally denoted as "T_g") characterizes amorphous/glass nature of materials. This is more easily understood in the case of a polymer. If we cool a polymer from its liquid state, initially it undergoes cooling and it gets a rubbery state and then after crossing the Tg, it becomes brittle. This kind of phenomenon occurs in amorphous metals too. In case of metallic glasses, we can say that T_g is the temperature at which material gets soft from hard upon heating or get hard upon cooling. This definition for polymers and metals looks similar but it is restricted to amorphous and semicrystalline metals only. The best way to explain the process of getting an amorphous metal or metallic glass is by supercooling the metal from its liquid state. In Figure T_f is the freezing temperature. During cooling, the liquid goes beyond the freezing point and is known as supercooled metal which can have an amorphous structure. In this way, we can get a metallic glass. In the absence of supercooling, the liquid has a tendency to crystallize.

During the formation of glass, the material should avoid the route of crystallization. Crystallization happens during the cooling of material below its liquidus temperature. The difference in Gibbs free energy between liquid and crystalline state is an important factor for the ability of a metal to crystallize or to become amorphous. Whenever there is a transformation between liquid to the solid-state of a material, the phase transformation at constant enthalpy gives a crystalline material. If enthalpy varies in that process, then the material escapes from the crystalline route and becomes amorphous. As we can see in Figure, a supercooling does not exist in the case of getting a crystalline material. On the other hand, in the case of supercooling, the enthalpy of transformation changes gradually. Therefore, during the manufacturing of the metallic glasses kinetics of the supercooling has a great impact on the quality of the glasses.

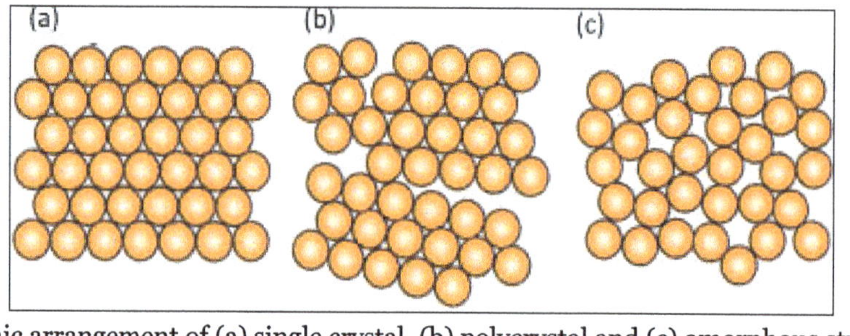

The atomic arrangement of (a) single crystal, (b) polycrystal and (c) amorphous structures.

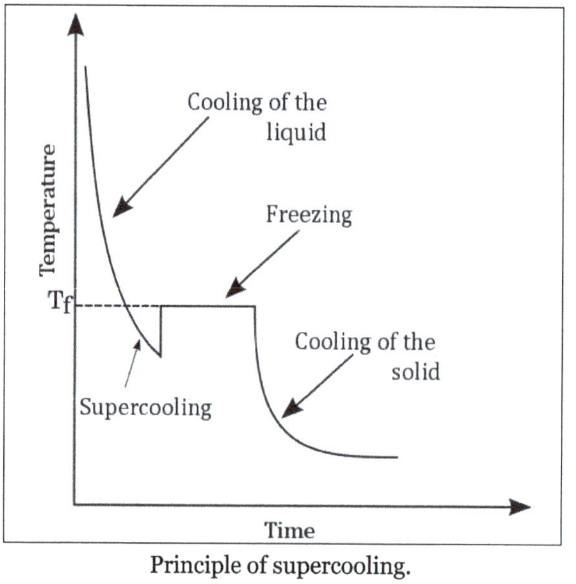

Principle of supercooling.

Considering the periodic table, the metallic glasses are mainly divided into two categories: metal–metal and metal–metalloid. In Figure elements which are metals and metalloids are shown in different colors for the better understanding of their selection for making metallic glasses.

As shown in the periodic table, metals are the elements starting from Lithium with atomic number 3. These are placed from group IA to VIA and shown by the yellow color in Figure. Some of the examples of metal–metal type metallic glasses are Ni–Nb, Mg–Zn, Hf–V, Cu–Zr, etc. Metals can be alkali, alkaline and rare earth metals etc. In metal–metal type, the atomic percentage of individual constituents can be up to 50%.

In the case of metal-metalloid type, one constituent is a metal and the other one is a metalloid. Metalloids like B, Si, Ge can be mixed with metals like Fe, Ni, Co etc. Metalloids in the periodic table are positioned in a step-like manner and colored with peach color. Metalloids have properties that are intermediate to both metals and non-metals. The composition percentage of metalloids in this category is lower than percentage of metals. After the discovery, the compositions of metalloids in the glasses were generally up to 20% but gradually people work on that problem and successfully decreased its percentage beyond 20%. Atomic size, the heat of mixing is also considered for making metallic glasses and used for classification of these materials. The properties based on base metal and categories of metallic glasses are provided in Tables, respectively. This type of glasses is more often used in commercial applications.

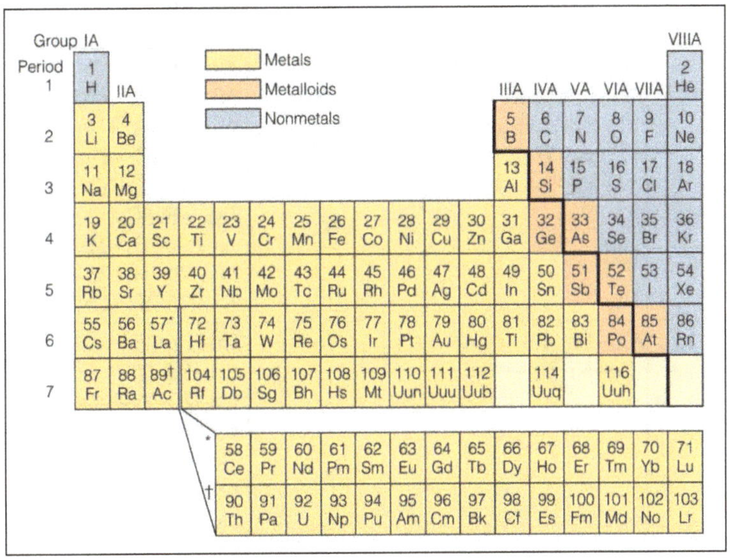

Periodic table showing metals, metalloids and non-metals.

Table: The basic properties of different LTM-based BMGs.

Base metal	Properties based on the base metal
Fe-based	Soft magnetism (glass, nanocrystal). Hard magnetism (nanocrystal). High corrosion resistance. High endurance against cycled impact deformation.
Co-based	Soft magnetism (glass, nanocrystal). Hard magnetism (nanocrystal). High corrosion resistance. High endurance against cycled impact deformation.

Ni-based	High strength, high ductility. High corrosion resistance. High hydrogen permeation.
Cu-based	High strength, high ductility (glass, nanocrystal). High fracture toughness, high fatigue strength. High corrosion resistanc.
Pd-based	High strength. High fatigue strength, high fracture toughness. High corrosion resistance.

Table: Composition and properties of different types of BMGs (composition of base metal is greater than 50 at. %).

Base metal	Metal–Metalloids	Metal–Metal
Fe-based	Fe–(Al,Ga)–(P,C,B,Si) Fe–Ga–(P,C,B,Si) Fe–Ga–(Nb,Cr,Mo)–(P,C,B) (Los Alamos) Fe–(Cr,Mo)–(B,C) Fe–(Zr,Hf,Nb,Ta)–B Fe–(B,Si)–Nb	Fe–Nd–Al
Co-based	Co–Ga–(Cr,Mo)–(P,C,B) Co–(Zr,Hf,Nb,Ta)–B Co–Ln–B	Co–Nd–Al Co–Sm–Al
Ni-based	Ni–(NbCr,Mo)–(P,B) Ni–(Ta,Cr,Mo)–(P,B) Ni–Zr–Ti–Sn–Si (Yonsei University) Ni–Pd–P	Ni–Nb–Ti Ni–Nb–Zr Ni–Nb–Hf Ni–Nb–Zr–Ti Ni–Nb–Zr–Ti–M (M = Fe, Co, Cu) Ni–Nb–Hf–Ti Ni–Nb–Hf–Ti–M Ni–Nb–Sn (Cal Tech)
Cu-based	Cu–Pd–P Cu–Ni–Pd–P	Cu–Zr–Ti Cu–Hf–Ti Cu–Zr–Ti–Ni Cu–Hf–Ti–Ni Cu–Zr–Ti–Y Cu–Hf–Ti–Y Cu–Zr–Ti–Be Cu–Hf–Ti–Be Cu–Zr–Al Cu–Hf–Al Cu–Zr–Al–M Cu–Hf–Al–M (M = Ni, Co, Pd, Ag) Cu–Zr–Ga Cu–Hf–Ga Cu–Zr–Ga–M Cu–Hf–Ga–M Cu–Zr–Al–Y (Cal Tech)
Pd-based	Pt–Cu–P Pt–Cu–Co–P (Cal Tech) Pt–Pd–Cu–P	

Structure, Properties and Applications

Structure of material defines its property. BMGs do not exhibit a long-range order structure, as they solidify from liquid without reaching the crystalline ground state. However, short to medium-range structural order does develop to a considerable extent under the given kinetic constraints. This happens because the atoms strive to find comfortable configurations to lower their energy. The structure of the bulk metallic liquids was first observed by Bernal and it was described as dense random packing. Structural features of metallic glasses are discussed by researcher where the concept of efficient filling of space is supported. The rationalization of the good glass forming compositions can be possible by the analysis of dense packing. An example of simple binary metallic glass is shown in Figure.

These structural motifs arise from the strong tendency to form as many bonds as possible between unlike species because of the large negative heat of mixing which is usual in good glass formers. The size of the cluster and its type depend on the relative size of the solvent and the solute. The replacement of Pt solute in Figure by much smaller Be reduce the number of Zr neighbors which can be accommodated around the solute, and the solute concentration in the alloy would be correspondingly much higher. The medium-range order and dense packing in three dimensional space can be possible by the overlapping of the cluster via various solvent-atom sharing schemes.

Model of a simple binary metallic glass: Interpenetrating quasi-equivalent clusters sharing faces, edges, or vertices in the atomic packing configuration of a Zr-Pt metallic glass. The blue balls represent the solvent Zr atoms centered around Pt solute atoms.

The adaptability of the metallic glass in the real world applications is spread in various fields, such as striking face plate in golf clubs, frame in tennis rackets, various shapes of optical mirrors, casing in cellular phones, casing in electromagnetic instruments, connecting part of optical fibers, shot penning balls, electromagnetic shielding plates, soft magnetic choke coils, soft magnetic high frequency power coils, high torque geared motor parts, high corrosion resistance coating plates, vessels for lead-free soldering, colliori type liquid flow meter, spring, in printing plate, high frequency type antenna material, biomedical instruments such as endoscope parts etc. Metallic glasses are very strong compared to other conventional materials and that makes it a very good candidate for military applications like armor (Bulletproof vest) piercing bullets, anti-tank projectiles etc. Those metallic glasses, which are stronger than titanium, are also tried for aerospace application. Recently researchers from NASA and different research centers and organizations of China, Britain and

Japan are doing several tests to get such ultimate material. This type of materials can give relatively double the performance compared to that of a titanium product in the space application. The major problem lies in the BMGs are the very quick aging of these materials. They become fragile after exposed to external physical stress conditions. The non-uniformity or not enough randomness inside the glass structure leads to the quick aging of these BMGs. So reliability cannot be achieved for the space applications. To remove such weakness of these materials, Wei-Hua Wang from China did experiments like the operation of a blacksmith. Cryogenic treatment of melted metal was done by using liquid nitrogen and maintained to room temperature after solidification. Again the material was melted and the process was repeated for several cycles. The purpose of choosing the repeated cryo-treatment is to reduce the instability of the material by increasing the randomness inside the material. This type of processing technique enhances the life of a BMG and increases its reliability.

Firstly China and secondly the United States are the major producers of BMGs. Currently, most applications are focused on electric based products like transformer core. Because of the good conductivity properties of BMGs, it dominates in that sector. Other applications like high-temperature applications, aerospace applications and military applications have a long way to go for becoming a better replacement for the recently used materials.

Processing of Metallic Glasses

The flow chart for the processing of metallic glasses in order to obtain different products is presented in figure.

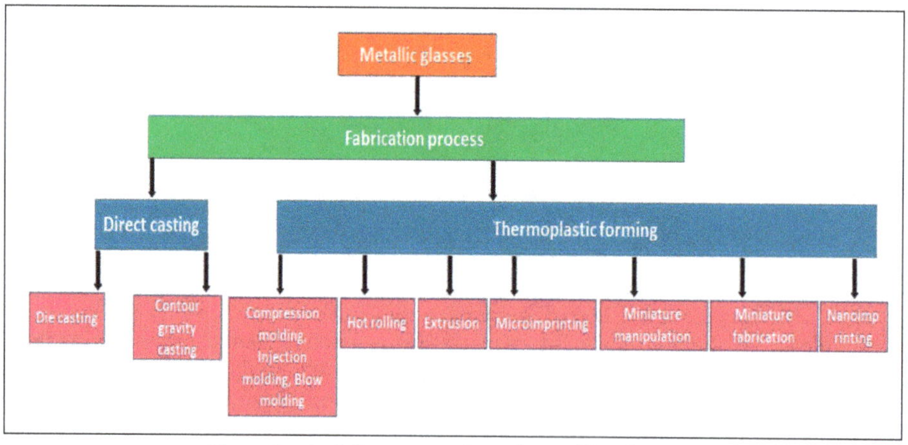

Flow chart of the processing of metallic glasses.

Direct Casting

For the net shape fabrication of BMG, two types of casting processes (suction and die casting) have been adopted. Among the two methods the suction casting method develops a product with higher quality and lower porosity than die casting method. BMGs having low melting temperature are beneficial because this process reduces tool cost and wear, lowers energy consumption and shortens cycle time. The TTT diagram and the critical cooling rate (T_c) are highly affected by the maximum temperature prior to cooling rate. The glass forming ability of the BMGs can be diminished if the maximum temperature prior to cooling does not exceed a threshold temperature (TOH). The overheating is rapid and it affects the viscosity in some type of BMGs. This viscosity effect can be broadly explained by the phenomenon including the melting of an oxide phase, ordering phenomenon and a chemical decomposition process. During the casting processes (both die casting and suction casting) shrinkage has to be taken in to consideration.

The shrinkage phenomenon is absent in the BMGs formers due to the absence of a first order phase transition during solidification. Cooling process in the casting plays a vital role. Low solidification shrinkage in BMGs develops a gap between the mold and the BMG during the cooling process. The heat transfer through the gap is different for the presence of atmosphere and vacuum, can become the rate limiting factor and leads to affect the cooling rate. In the process of direct casting of BMG formers, fast cooling and forming have to be done simultaneously due to the crystallization mechanism and the crystallization kinetics. Direct casting needs care during the mold filling and at the same time to avoid crystallization during solidification. This process is very difficult while making an intricate part which has attractive properties. Figure represents some application of BMGs fabricated by direct casting. The advantages and disadvantages of the direct casting BMG process are given in table.

Table: Advantages and disadvantages of the direct casting process.

Advantages	Disadvantage
Low melting temperature	Cooling and forming are coupled
Low shrinkage	Processing environment can influence crystallization kinetics
One step process	High viscosity
Homogeneous microstructure	Internal stresses
Mechanical properties are already matured in the as-cast state	BMGs contaminate during processing

Thermoplastic Forming

Thermoplastic forming (TPF) is the alternative process to direct casting for developing BMGs. This process has different nomenclature such as hot forming, hot pressing, superplastic forming, viscous flow working and viscous flow forming. The preferable condition for TPF is the drastic softening of the BMG former upon heating above Tg and its thermal stability. The measure of the ability of BMG formers to adopt an amorphous structure by heating above its glass transition temperature is known as thermal stability and it can be quantified by the width of the supercooled liquid region (SCLR). For $Zr_{44}Ti_{11}Cu_{10}Ni_{10}Be_{25}$ a very long processing time is available at low temperature by a high viscosity. Furthermore, the viscosity is significantly reduced at a high temperature which leads to the reduction in processing time. Low viscosity and the long processing time are the optimum parameters for the highest formability of BMG former in its SCLR. The thermos physical properties that are detecting good formability of a BMG former are fragile liquid behavior, large poison ratio and low glass transition. Some TPF based BMG formers the glass transition temperature and the softening in the SCLR permits the use of processing temperature and pressure. The capacity to plastically form metallic glasses in the zone of T_g was recognized by researchers. The enhanced formability of BMGs, which is developed by a wide range of processing methods based on the thermoplastic forming, attracts a wide range of researchers. In all the above-said processes, one thing is common that is the "feedstock". The shape and size of the feedstock are different for powder, over rods, disks, plates and films.

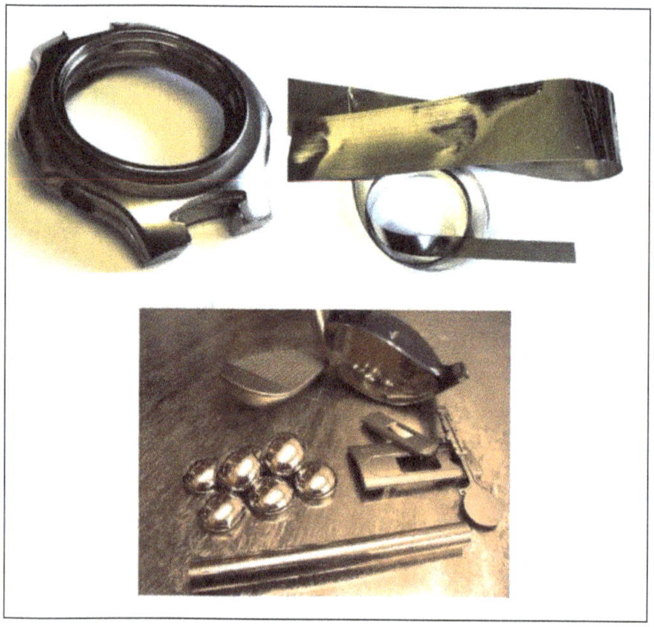

BMG articles fabricated by direct casting method.

TPF based Compression and Injection Molding

Compression molding was an adopted form of plastic processing. In this process, the feedstock material is placed in a mold and is given temperature into SCLR and pressure must exceed the flow stress of the BMG to achieve the required strain prior to crystallization setting. Fast cooling is not required for the forming. Figure represents the schematic diagram of the process equipped with examples. Under low applied pressure in the compression molding Pd-Ni-Cu-P alloy can be formed as a gear-shaped structure. It can be noticed that a dense compact part with mechanical properties close to those of bulk material and an outstanding surface finish can be obtained. Figure indicates compression molding of BM6 former, having different feedstock shapes such as pellets, plates and rods. The needed molding pressure not only depends on the formability of the BMG at the processing temperature but also on the shape of the final product. Injection molding is also a TPF based molding process. The only difference between the injection molding and compression molding is that in injection molding, feedstock material is rendered into the mold cavity which has the benefits for the development of the commercial fabrication processes with minimized cycle time.

Miniature Fabrication

The development in technologies like micro-electromechanical systems (MEMS), electronics devices, and medical devices have created a rising demand for miniature products and parts. The miniature formation is done by different processes like German method LIGA (lithography, electroplating and molding), UVLIGA etc. Due to the drawbacks of the LIGA i.e. cost, UV-LIGA is developed giving the similar products. Both processes can be used as the surface patterning techniques for creating high aspect ratio structures. Due to the homogeneous and isotropic structure of BMGs in atomic scale and their superior properties over conventional materials used for miniature applications and the capability to produce stress-free parts, these methods attracted a lot of attention.

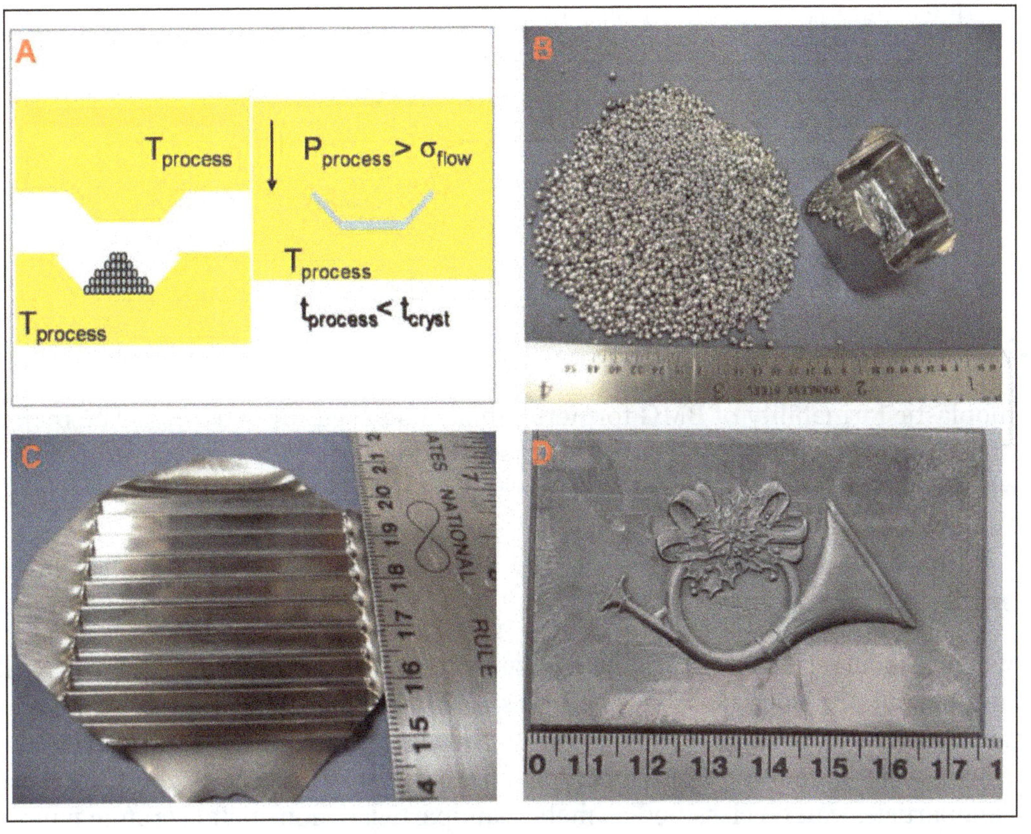

TPF based compression molding with BMG. (A) Schematic diagram of the compression molding with BMGs. (B) Pellets used as feedstock material to compression mold $Pt_{57.5}Cu_{14.7}Ni_{5.3}P_{22.5}$. ($Zr_{44}Ti_{11}Cu_{10}Ni_{10}Be_{25}$) formed from a flat plate into a corrugated structure. (D) $Zr_{44}Ti_{11}Cu_{10}Ni_{10}Be_{25}$ formed from a flat plate to create an embossing mold.

Nano Forming

BMGs basically formed by top-down nanofabrication. The combination of different properties like high strength at room temperature, the ability to imprint a nanometer-sized parallel print process, the non-linear softening of BMGs when reaching their glass transition and the ability to repeatedly write and erase facility on the BMG surface recommends a wide range of application. Nanoimprinting on BMG permits to directly write, such as with atomic force microscopy (AFM) tip as in a scanning probe lithography process. The capability of BMGs for direct nanoimprinting can be applied with a combination of surface smoothening method and used as a rewritable high-density data storage. Several materials have been developed for the mold formation and imprint for nanoforming such as silicon, quartz, and alumina.

Rolling

Rolling of metallic glasses can be categorized into two processes; one is based on liquid processing and the other is on thermoplastic forming. The example of the former is the melt spinning. In the melt spinning, the liquid sample is quenched by injecting on a single, fast-spinning copper roll. Another process is cold rolling where the BMGs are rolled at the room temperature.

Table: Advantage and disadvantages of TPF based BMG processing.

Advantages	Disadvantages
Forming and fast cooling decoupled	Two or more step process
Highest dimensional accuracy	
Insensitive to heterogeneous influence	Novel and unique process
Novel and unique process "green" process	
Low capital investment	

Extrusion

The thermoplastic formability of BMG formers can also be used for extrusion. The major advantage of extrusion is that it produces the highest aspect ratio shapes, 100+, of uniform cross-section. During extrusion after the outlet of the effective die length, the swelling of the material is a common phenomenon.

Nanocomposite Materials

The nanoparticle includes the particles having size between 1 and 100 nm. These particles have different properties at their atomic level due to their size. This change in properties of nanoparticles is beneficial in many fields. Nanotechnology is one of the most interesting fields for researchers since the last century. Numbers of developments have been made since then in the field of nanotechnology. Nanoparticles can be classified as metal nanoparticles, non-metal ceramic nanoparticles, semiconductor nanoparticles, and a well-known type is carbon nanoparticles. Nanoparticles have those chemical and physical properties which makes them very different from that of the corresponding bulk materials due to their small size and large surface to volume ratio. They attract much attention because of their potential applications in many fields including optics, electrics, magnetism, ceramics, and catalysis.

Composites

Composites are engineered or naturally occurring solid materials which results when two or more different constituent materials, each having its own significant characteristic (physical or chemical properties) are combined together to create a new substance with superior properties than original materials in a specific finished structure. They are commonly designed to offer wide range of properties and characteristics, some are as follows:

- Stiffness and strength.
- Low coefficient of expansion.
- Resistance against fatigue.
- Ease in manufacturing complex shapes.
- Simple repair of damaged structures.
- Resistance to corrosion.

Nanocomposites

Nanocomposites are those composites in which one phase has nanoscale morphology like nanoparticles, nanotubes, or lamellar nanostructure. They have multiphases, so are multiphasic materials, at least of the phases should have dimensions in the range of 10–100 nm. To overcome the limitation of different engineering materials now-a-days, nanocomposites are emerged to provide beneficial alternatives. Nanocomposites can be classified on the basis of their dispersed matrix and dispersed phase materials. With the help of this rapidly expanding field, now-a-days, it has been possible to generate many exciting new materials with novel properties via innovative synthetic approaches. The properties of the so-called found not only depended on the properties of their originals, but also crucially on their interfacial and morphological characteristics. Of course, we cannot ignore the fact that sometimes it also happened that the newly generated property in the material is unknown to the parent constituent materials. Hence, the idea behind nanocomposite is to use building blocks with dimensions in nanometer range to design and create new materials with unprecedented flexibility and improvement in their physical properties.

Advantages of Designing Novel Nanocomposites

Nanocomposites are the solid combination of a bulk matrix and nanodimensional phase(s) which differ in properties due to dissimilarities in structure and chemistry. Properties that have indicated substantial improvements:

- Mechanical properties (strength, bulk modules, withstands limit, etc.)
- Thermal stability.
- Hinders flame and reduce smoke generations.
- Permeability of gases, water, and solvents are reduced.
- More surface appearance.
- Improved electrical conductivity.
- Increased chemical resistance.
- Enhance optical clarity as compared to conventionally filled polymers.

Among several nanocomposites, polymer-based nanomaterials are the most leading materials of current research and development. Characteristics like film forming ability, activated functionalities, and dimensional variability provide lots of benefit to polymer-based nanocomposites.

The potential risk of nanocomposites commonly occurs majorly in areas like:

- Risk to health and environment.
- Molecular manufacturing.
- Societal risks.

Formation of Nanocomposite Materials

Nanocomposites can be formed by blending inorganic nanoclusters, fullerenes, clays, metals, oxides, or semiconductors with numerous organic polymers or organic and organometallic compounds, biological molecules, enzymes, and sol-gel derived polymers.

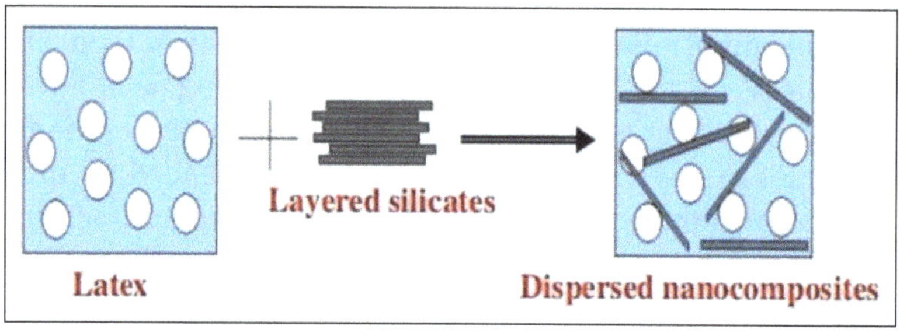

Formation of nanocomposite materials.

Nanocomposite materials that are obtained by the combination of two or more separate building constituents in one material offers unique properties that plausibly arises from their small size, large surface area, and off course from the interfacial interaction between the phases. Their extra ordinary potential have been smoothly utilized to enhance the biological potential of many drugs, biomaterials, catalysts, and also in some high-value added materials.

Types of Nanocomposites

Nanocomposite materials can be classified in the following way based on the presence or absence of polymeric material in the composite. The nanocomposites in which the compositions do not contain any polymers or polymer-derived materials are called non-polymer-based nanocomposites. Non-polymer-based nanocomposites are also known as inorganic nanocomposites. They can be further classified into metal-based nanocomposites, ceramic-based nanocomposites, and ceramic-ceramic-based nanocomposites.

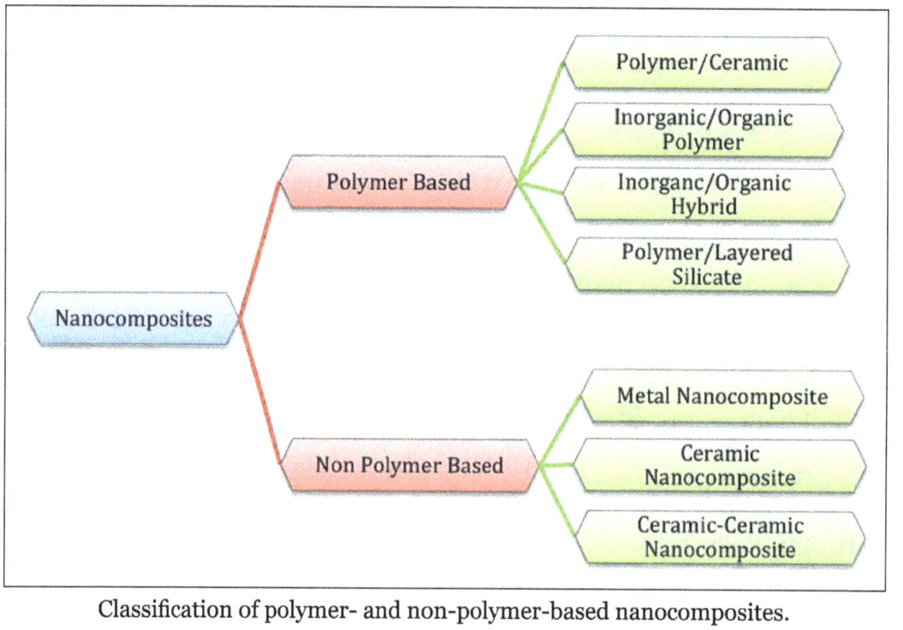

Classification of polymer- and non-polymer-based nanocomposites.

Non-Polymer-Based Nanocomposites

Metal-Based Nanocomposites

Bimetallic nanoparticles being investigated in detail in the form of either of alloy or core-shell structures due to their improved catalytic properties and advancement in optical properties related to individual and differentiate metals. They can be characterized by:

- Super plasticity,
- Lower melting points,
- Increased strength and hardness,
- Improved magnetic properties,
- Increased electrical resistivity, etc.

Non-polymer-based nanocomposites can be also classified as: metal/metal nanocomposites, for example Pt-Ru nanocomposites.

Ceramic-Based Nanocomposites

Ceramic-based nanocomposites are defined as ceramic composites with more than one solid phase, in which at least one of the phases has dimensions in the nanoscale range (<50–100 nm). In these types of composites, both the phases have combined magnetic, chemical, optical, and mechanical properties, for example hydroxyapatite/titania nanocomposites.

These can be characterized by:

- Better toughness.
- Increased ductility.
- Increased strength and hardness.

Ceramic/Ceramic-Based Nanocomposites

The non-polymer-based nanocomposites can be also classified as ceramic/ceramic nanocomposites which can be used in the area of artificial joint implants for fracture failures and it could promptly reduce the cost of surgery and would extend the mobility of the patient. The life spam would increase by 30 years, if the use of zirconia-toughened alumina nanocomposite implants is used effectively. The other example of ceramic/ceramic nanocomposites are calcium sulfate-biomimetic apatite nanocomposites.

The most promising prospects of both metal-based nanocomposites and ceramic-based nanocomposites are in the application of areas in dentistry in which the non-polymer-based nanocomposites or the inorganic materials that is metal or ceramics such as calcium phosphate, hydroxyapatite, and bioactive glass nanoparticles are very advantageous in alveolar bone regeneration and enamel substitution.

Polymer-Based Nanocomposites

The polymer or copolymer which contains nanoparticles or nanofillers dispersed in the polymer matrix is termed as poly nanocomposites. One dimension (1D) must be lying in the range of 1–50 nm and these possess several shapes like as platelets, fibers, spheroids, etc. Poly nanocomposites are in the category of multiphase systems such as, MPS namely composites, blends, and foams which can absorb about 95% of the production of plastics. So, these systems need controlled mixing, the achieved dispersion should be stable, dispersed phase should be oriented, and the compounding strategies which are involved for all MPS, which includes poly nanocomposites (PNC) is almost same.

Polymer nanocomposites are proposed as a class of materials with unique properties but, the most challenging property of PNCs is the complex interfacial areas in between the polymer matrices because of this small scale large specific area is created that highlight the importance of polymer-nanoparticle interactions. So, to achieve properties, such as, mechanical, thermal, optical, and electric, we need to analyze the intercalation process among the nanoparticles and polymer bases.

Polymer nanocomposites are known to be a class of reinforced polymer with a very low, i.e., less than about 5% of nanometric clay particles. These substances gained huge attention simultaneously from both the academic institution as well as from industrial sectors commonly in the area of nanocomposites. This is actually due to their drastically enhanced or improved thermal, mechanical as well as the barrier properties as compared to the micro- and also the conventional composites. These materials can be differentiated notably by: improved fire resistance and thermal stability, improved barrier properties, and increased recyclability.

However, despite of having so many advantages, it is still very much difficult to prepare a uniform dispersion between the filler and the matrix, as shown in Figure. Hence, unlikely, it reflects the lower mechanical as well as thermal properties in the produced nanocomposites.

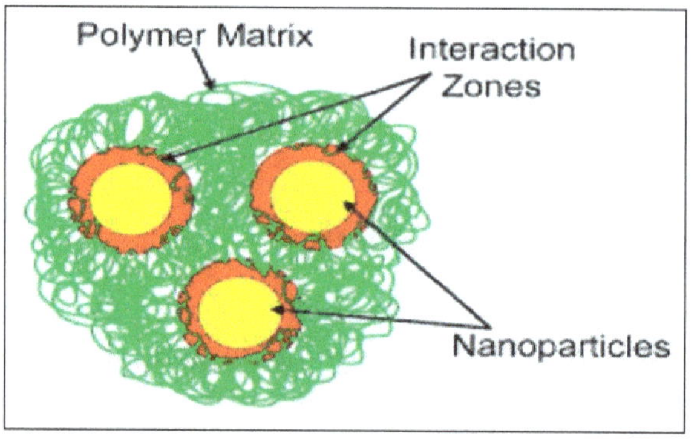

Uniform dispersion between the filler and matrix in nanocomposites.

Uses of Polymer Nanocomposites

Figure shows the various uses of polymer nanocomposites irrespective of the nature of the field used. By the hydrolysis of tetraethyl-ortho-silicate, the hybrids made of poly rubber (dimethyl siloxane) and nanosilica can be given a specific shape like objects, such as golf balls. Many number of polymer nanocomposites for example, rubber, propylene, styrene butadiene rubber, and ethylene vinyl acetate are used in barrier applications.

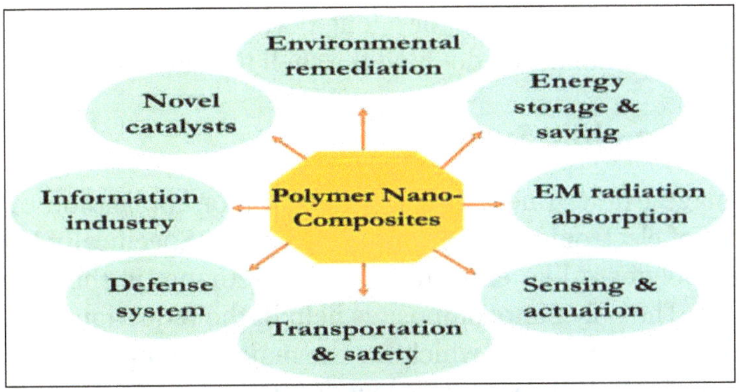

Various uses of polymer nanocomposites.

Rubbery hybrids with different shapes.

They can act as a tremendous barrier for chemicals like toluene, sulfuric acid, and hydrochloric acid as well as for several gases such as, carbon dioxide, oxygen, and nitrogen. They are also utilized in chemical protective and surgical gloves as they have excellent solvent barrier properties in order for avoiding contamination from medicine.

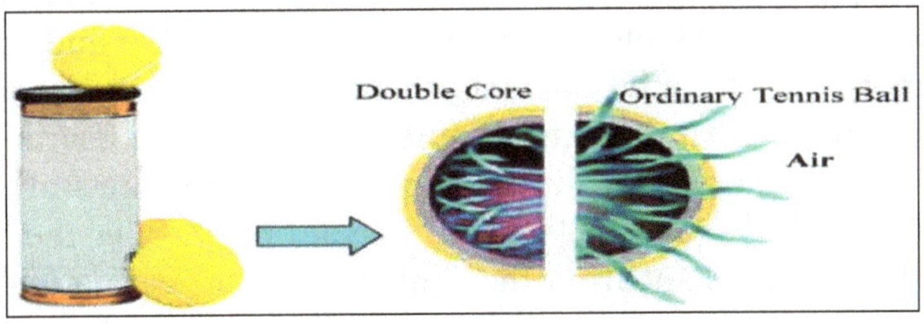

The core of this Wilson tennis ball is covered by a polymer-clay nanocomposite coating.

Polymer nanocomposites are also used in food packaging, and the particular examples for food packaging includes processed cheese, meat, and dairy products also the medical containers for carrying blood collection tubes, baby pacifiers, and drinking water bottles. To enhance the barrier, mechanical properties and the life of the product clay-based polymer nanocomposites are been used in plastic bottles. Nanocomposites are also incorporated for beer bottle manufacture, so as to reduce many problems like biological and non-biological aspects, beer colloids instability, oxygen permeation, and change in taste due to light exposure. The double core Wilson tennis ball is the most recently commercialized sports goods, in which the coating of clay nanocomposites is done in order to maintain the internal pressure for a long period of time and also the core is coated by butyl rubber clay nanocomposite which doubled the shell life and acts a gas barrier.

Electronics and Automobile Sectors

In today's time, the biggest milestone is the incorporation or application of polymer/clay nanocomposites in the field of electronics and automobile sectors. Specifically to decrease the solvent transmission through polymers like elastomers, poly urethane, is the most impressive ability nanoclay incorporation. The poly nanocomposites help in the reduction of weight and processing cost so that they used by tire companies which are the major driving force for their usage. It is the naturally obtained materials which possesses low density. When the clay incorporated tires are compared with the ordinary ones then it is found that they have excellent mechanical properties and also improved gas barrier performance for tubeless tires uses. Mostly for automobile tire manufacturing, styrene butadiene and natural rubber nanocomposites are preferred the most. It is due to their improved thermal properties and abrasion resistance that contribute to the long life of the tires. They have great applications in solar cells, transistors, battery manufacturing, etc.

Coatings

The most important modifying property of surfaces is coating. So, many methods and strategies are tried by the researchers to improve the surface properties of several products. The properties such as, excellent resistance for chemicals, better barrier properties, super hydrophobicity, and corrosion resistance are exhibited by nanoclay incorporated thermoset polymer nanocoatings. The process parameters, such as dipping time, nature of surfactant, temperature, etc., determine the coating thickness. The thermoset polymer nanocoatings which are incorporated clay and nanosilver could improve the antibacterial properties and is used in medical sectors.

Bio-Nanocomposites

The name "Bi-nanocomposites" is given as they are characterized as natural nanocomposite. To understand their essential roles in biological systems, their structures and properties are studied by biologists. Bio-nanocomposites are designed originally and are present to fulfill the needs of life and to meet surrounding environmental conditions so they can guarantee the living of the associated species. Natural materials are different in terms of structure and compositions but the design of bio-nanocomposites require biological molecules to consider them as synthetic building blocks, which is far more distant from the context of their own natural function. They are made of biopolymers and inorganic solids which has the dimension in the range of 1–100 nm. Due to their multidimensional properties such as antimicrobial activity, biocompatibility, and biodegradability they have several numbers of applications. The effective outcome of growing needs of bio-based polymers is the drastic reduction in the usage of fossil fuels. Bio-nanocomposites have easily replaced conventional non-biodegradable petroleum-based plastic as they are light weight and eco-friendly; they have become a sustainable that is future lasting material for use in high performance applications. As they are biocompatible, it makes them beneficial for biomedical applications and also makes them suitable for cosmetics and biotechnology applications. They have dominant significance in the future as green sustainable materials. Bio-nanocomposites will act as substituents for the currently used petroleum-based polymers.

Uses of Bio-Nanocomposites

- Bio-nanocomposites are used in cosmetics industries and also in the fabrication of implants, scaffolds, diagnostics, and biomedical devices.
- They are also very beneficial as catalysts, contact lenses, and gas-separation membranes.
- In the treatment of osteomyelitis, by the regeneration of tissue biopolymer-based nanocomposite is applicable.
- Artificial bone implants involves nanostructured organic/inorganic nanocomposites which are useful in managing load-bearing bone grafts.
- Using live cells of functionalized particles, controlled electrophoretic assembly of bioinorganic composite materials is done.
- They are largely applicable in diagnostic, drug delivery, and tissue generation.
- In industries, they are used as actuators. They are also used in the manufacturing of compostable bags as they are eco-friendly.

Applications of Nanocomposites

Nanocomposites have been growing with a speedy rate so as their large number of applications. In the next 10 years, the worldwide production will exceed 600,000 tons in the following regions:

- Superior strength fibers and films.
- UV protection gels.
- Drug delivery systems.
- New fire retardant materials.
- Anti-corrosion barrier coatings.
- Lubricant and stretch paints.

Nanocomposites have also attracted the field of automotive and industrial applications by doing enhancements in especially the mechanical properties. They can be used or applied in the various vehicles types like engine covers, door covers, and timing belt covers. Other applications are usage as blades for vacuum cleaners, mower hood, covers for mobile phones, etc.

Future Aspects

Modification of surface properties of nanoparticles by treating them with green agents for specific applications having specific improved microstructural properties like improved exfoliation, compatibility, and also thermal stability.

Advance optimization of the polymerization conditions during the preparation of the nanocomposites in order to get maximum output with minimal cost. Detailed study on the effect of composition of the nanocomposites to build up the developed microstructures during the preparation activities.

Preparation of nanocomposites as well as their blends by using the materials like polymer blends along with the melt blending technologies. Hence, the advantages of the properties of the individual material as well as their coaction can be developed.

Using nanocomposites to make flexible batteries: "A nanocomposite of cellulous materials and nanotubes could be used to make a conductive paper. When this conductive paper is soaked in an electrolyte, a flexible battery is formed."

References

- What-is-materials-science, resources: materialseducation.org, Retrieved 29, June 2020
- What-is-material-definition-of-materials, nuclear-engineering, materials-science: nuclear-power.net, Retrieved 07, May 2020
- An-introduction-to-biomaterials: aladdin-e.com, Retrieved 20, January 2020
- Metallic-Glasses-A-Revolution-in-Material-Science-339062814: researchgate.net, Retrieved 15, July 2020
- Nanocomposite-materials, nanotechnology-and-the-environment: intechopen.com, Retrieved 07, March 2020

2

Characterization Techniques of Materials

There are various characterization techniques of materials like scanning electron microscopy, X-ray photoelectron spectroscopy, focused ion beam, X-ray diffraction, transmission electron microscopy and nanoindentation etc. This chapter has been carefully written to provide an easy understanding of the various characterization techniques of materials.

Thermogravimetric Analysis

TGA (Thermo Gravimetric Analysis) is an important characterization technique which measures change in mass of a material with variations in temperature. It operates in two modes: 1) change in mass is measured with increasing temperature (called scanning mode having constant heating rate); and 2) change in mass is measured with time (called the isothermal mode with constant temperature or constant mass loss). This change of mass is due to both physical and chemical phenomenon like absorption, sublimation, vaporisation, oxidation, reduction and decomposition likewise.

Types of TGA

- Dynamic TGA: In case of dynamic TGA, specimen is heated with a constant heating rate (i.e. normally linear increase with time).
- Isothermal or static TGA: In case of isothermal TGA, specimen is kept at a constant temperature for predefined time and the variations in weight (or mass) of the specimen are monitored.
- Quasi static TGA: In this method, the specimen is heated to a constant weight at each step of the series of increasing temperature.

Principle

Microbalance records the change in weight of a specimen against the temperature or time, in specific controlled conditions including rate of heating, gaseous medium, flow rate, and crucible type, etc. Graph of change in weight with temperature is called thermo gravimetric curve or thermo gram.

Instrumentation

Medium TGA equipment constitutes a highly sensitive balance (a microbalance) for continuous measurement of specimen's weight, a furnace enclosing the sample holder, and a purged gas system to supply gas (inert or reactive, as per the requirement). The furnace is controlled by a computer which also collects the measurement data (sample temperature v/s weight). The balance is calibrated in a manner that a change in weight of the sample produces a proportional electrical signal. A data processing system then collects this electrical signal and converts it into weight or weight loss, which is then plotted on the y axis of the thermal curve. Block diagram of a typical thermos gravimetric analysis setup is shown in Figure.

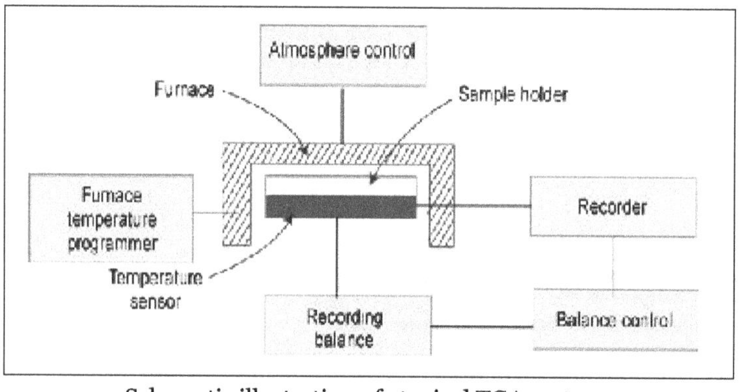

Schematic illustration of atypical TGA system.

There are two possible configuration modes of balance and furnace i.e. vertical furnace and horizontal furnace and both of them suffer from drift with the increase in temperature. Vertical configuration suffers with buoyancy effect attributable to a variation in the density of gas with temperature. Whereas, the horizontal arrangement, which is devised to reduce these buoyancy effects, experiences change in quartz rod length (this rod links the sample to the balance) as the temperature varies. Buoyancy effect is due to the force exerted by fluid that the opposes the weight of an increased object Modern instruments that heat at rates of 1000 °C/min are available and commercial instruments that heat at rates of 200 °C/min from room temperature to about 1200 °C can also be designed. In these systems, cooling by forced air can be done at 50 °C/min.

Sample Preparation

Maximizing the sample surface area in a TGA pan enhances the resolution and reproducibility of weight loss with temperature. The sample weight influences the accuracy of weight loss measurements. Usually 10-20 mg of material is desirable in most applications and for volatile materials 20-100 mg of sample is required. The instruments have a baseline drift of +-0.25% of a 10 mg sample.

Calibration

Calibration of mass is performed using standard mass at room temperature: removing standard weight on sample holder, where in recorded mass change is calibrated to the mass of standard weight. Heat induced drift in the baseline is monitored by measurement using blank sample pan. After these measurements, the standard sample is subjected to similar measurement parameters, that is, heating rate, gas flow rate, type of gas used, final temperature, etc. The standard sample experiences quantitative decomposition during the reaction steps. Mass change correction is made by subtracting the apparent mass as a function of temperature of sample being investigated. Mass loss corrected experimentally should be confirmed and should be in good agreement with the calculated one.

Kinetic Theory

The transformation rate, dα/dt, varies linearly with temperature-independent conversion function, f(α) and the temperature-dependent constant, k(T), for solid-state heterogeneous reactions; and is expressed as:

$$\frac{d\alpha}{dt} = k(T)f(\alpha)$$

Rate constant temperature dependence is usually explained by the Arrhenius equation. Explanation of the data mathematically is described via three kinetic parameters: activation energy E, pre exponential factor A, and the kinetic model algebraic expression in conversion function α(f (α)).

$$\frac{d\alpha}{dt} = Ae^{\frac{E}{RT}}(\alpha)$$

At the same time,

$$\beta = \frac{dT}{dt} = \text{Constant}$$

Here β is heating rate, By using value of β from Equation $\beta = \frac{dT}{dt} = \text{Constant}$ in Equation $\frac{d\alpha}{dt} = Ae^{\frac{E}{RT}}(\alpha)$, the above rate expression is changed to the expression for non-isothermal rates, which describe the reaction rate as a function of temperature at fixed β values. Equation $\frac{d\alpha}{dt} = Ae^{\frac{E}{RT}}(\alpha)$ changes to:

$$\frac{d\alpha}{dT} = \frac{1}{\beta}Ae^{\frac{E}{RT}}f(\alpha)$$

Integrating Equation above up to transformation α directly yields:

$$\int_{o}^{\alpha}\frac{d\alpha}{f(\alpha)} = g(\alpha) = \frac{A}{\beta}\int_{T_o}^{T} e^{\frac{-E}{RT}}dT$$

According to non-isothermal iso-conversional methods applied by Ozawa, Wall and Flynn using the Doyle's approximation of p(x), Energy of activation can be determined by assessing the temperatures corresponding to fixed values from experiments at different heating rates:

Activation energy can be evaluated experimentally by calculating temperatures at specific values corresponding to variable heating rates as:

$$\text{In}(\beta) = \text{In}\left[\frac{AE}{Rg(\alpha)}\right] - 5331 - 1052\frac{E}{RT}$$

Thus, the plot of ln β versus 1/T yields the activation energy. The order of the reaction can be determined from the variation of the degree of conversion with temperature and heating rate expressed as:

$$\alpha(T) = 1 - e\left[\frac{kT}{\beta n}\right]$$

Taking double logarithm of both sides of Equation above with substituting Arrhenius equation gives.

$$\text{In}\left[-\text{In}\left(1-\alpha(T)\right)\right] = \text{In } A - \frac{E}{RT} - n / \text{In } \beta$$

A plot of ln[-ln(1-α(T)] versus ln β, which is obtained at constant temperature from different isotherms taken at different rate of heating, yields lines. The slope of the plot gives reaction order, which is also termed as Flynn–Wall–Ozawa exponent 'n'.

Example of TGA Curve

Heat Decomposition Mass Curve of Whewellite (Calcium Oxalate Monohydrate) is shown in Figure.

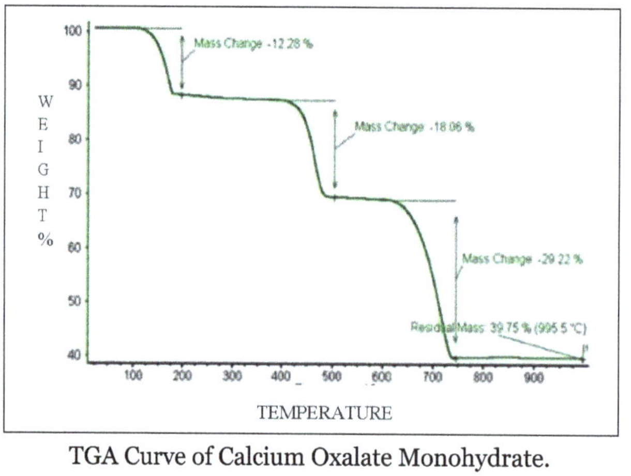

TGA Curve of Calcium Oxalate Monohydrate.

The successive plateaus signify the development of anhydrous salt, calcium carbonate and calcium oxide, respectively, and can be given by following chemistries:

- $CaC_2O_4.H_2O \rightarrow CaC_2O_4 + H_2O$
- $CaC_2O_4 \rightarrow CaCO_3 + CO$
- $CaCO_3 \rightarrow CaO + CO_2$.

The thermo gravimetric curve signifies that removal of water starts at 100 °C and removal of CO at 450 °C and CO_2 at 600 °C.

TG Curves

The plot of weight/mass versus temperature or time created by a thermogravimeter is termed asthermogram. There are seven different types of TG curve obtained from TGA as shown in Figure.

1. There is no decomposition with removal of volatile products.

- Fast initial weight loss. It is characteristic of sample drying (repeat experiment after drying the sample).
- Single phase decomposition.
- Multi-step decomposition with stable intermediate products.
- Multi-step decomposition with no stable intermediary. Nevertheless heating-rate may affect the plot and therefore, must be taken into account. At slow heating rates, type (5) resembles type (4) curve.
- Certain reactions (such as oxidation of metals) with atmosphere may result in increase in mass.

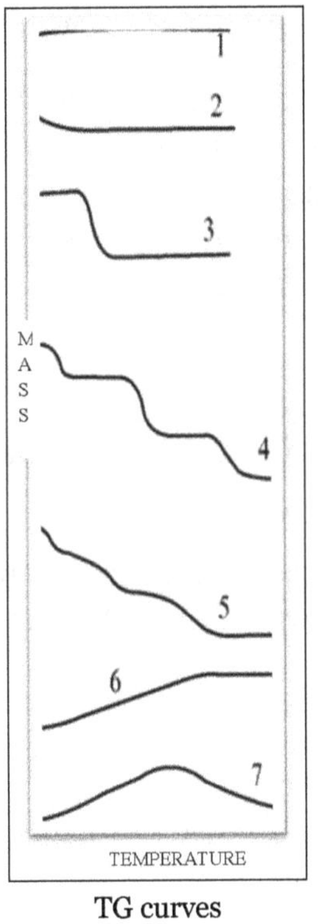

TG curves

Formalized TG Curve

- Plateau: A plateau represents a section on the TG curve in which there is no change in mass or the mass is actually constant. This has been shown by 'AB' in Figure.
- Procedural decomposition temperature: It is the temperature(in degrees C or K) at which the cumulative change in massattains a magnitude which can be detected by the thermo balance. This has been shown by 'B' in Figure.

- Final temperature: It is the temperature at which the cumulative mass change reaches its maximum value.

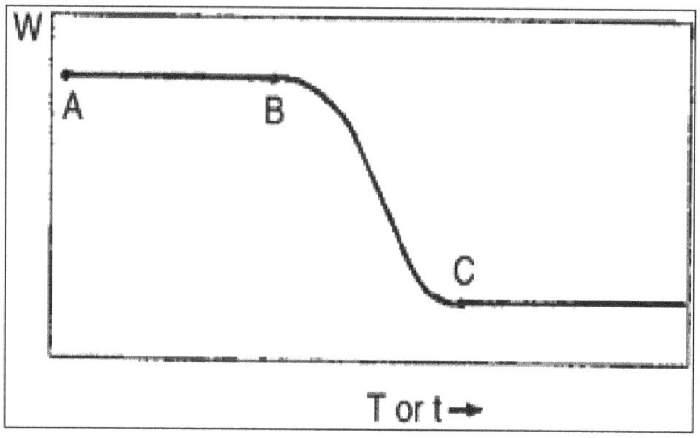

Formalized TG curve.

Information from TG Curve

TG curves can be analysed to obtain the following information:

- Plateau: It indicates the region of thermal stability of the sample/substance in which there is no mass change in sample. This ability of sample to retain its properties nearly unchanged on heating is called thermal stability. By knowing the thermal stability of the sample, engineers can find out range of temperature in which substance likes polymers metals alloys building materials can be used without any compositional breakdown.
- Weight/mass loss: When a sample is heated at a particular temperature, the sample losses its weight/mass. To determine the composition of sample/substance and find out the reactions involved during decomposition, inorganic chemists often use weight /mass loss during the decomposition.
- Procedural decomposition temperature: Procedural decomposition temperature can be determined from thermogram which indicates the evaporation or decomposition of substance /sample.
- Composition: Heating of sample leads to mass /weight loss which helps to determine the constituents of sample/substance. This information can be used to determine the quantity (in percentage) of a particular substance in a mixture.

There are two factors which affects the TG curve:

- Instrumental factors: It depends on furnace environment, gas flow rate, heating rate, geometry and material of the pan. Normally heating rate of 3.5 °C per minute is considered suitable for reliability and reproducibility of TGA.
- Sample characteristics: It depends on mass (small mass of sample is recommended), particle size (fine powder is suitable), sample preparation or pre-treatment, packing (compact solids), heat of reaction, thermal conductivity and sample purity.

There are several processes that occur in TGA which leads to weight loss and gain.

Table: Process that leads to weight loss and gain in TGA.

Process	Loss	Gain
Adsorption or absorption		✓
Solid gas reactions	✓	✓
Magnetic transitions	✓	✓
Desorption, Drying	✓	
Dehydration, Desolvation	✓	
Sublimation	✓	
Vaporization	✓	
Solid state reactions	✓	

Applications

Pharma Materials

DSC and TGA are often used for characterisation of pharma materials. DSC is able to differentiate between different polymorphic structures and, by using different heating rates, can investigate the transformations which occur during the polymorphic transformation. By using appropriate heating rates, polymorphic purity can be determined, and can involve heating rates up to 750 °C/min. TGA is often used to measure residual solvents and moisture, but can also be used to determine solubility of pharma materials in solvents. Analysis of pharma materials is probably the largest area of application for thermal analysis.

Polymers

Polymers represent another large area in which thermal analysis finds strong applications. Thermoplastic polymers are commonly found in everyday packaging and household items, but for the analysis of the raw materials, effects of the many additive used (including stabilisers and colours) and fine-tuning of the moulding or extrusion processing used can be achieved by using DSC. An example is oxidation induction time (OIT) by DSC which can determine the amount of oxidation stabiliser present in a thermoplastic (usually a polyolefin) polymer material. Compositional analysis is often made using TGA, which can separate fillers, polymer resin and other additives. TGA can also give an indication of thermal stability and the effects of additives such as flame retardants

Thermal analysis of composite materials, such as carbon fibre composites or glass epoxy composites are often carried out using DMA which can measure the stiffness of materials by determining the modulus and damping (energy absorbing) properties of the material. Aerospace companies often employ these analysers in routine quality control to ensure that products being manufactured meet the required strength specifications. Formula 1 racing car manufacturers also have similar requirements! DSC is used to determine the curing properties of the resins used in composite materials, and can also confirm whether a resin can be cured and how much heat is evolved during that process. Application of predictive kinetics analysis can help to fine-tune manufacturing processes. Another example is that TGA can be used to measure the fibre content of composites by heating a sample to remove the resin by application of heat and then determining the mass remaining.

Metals

Production of many metals (cast iron, grey iron, ductile iron, compacted graphite iron, 3000 series aluminium alloys, copper alloys, silver, and complex steels) are aided by a production technique also referred to as thermal analysis. A sample of liquid metal is removed from the furnace or ladle and poured into a sample cup with a thermocouple embedded in it. The temperature is then monitored, and the phase diagram arrests (liquidus, eutectic, and solidus) are noted. From this information chemical composition based on the phase diagram can be calculated, or the crystalline structure of the cast sample can be estimated. Strictly speaking these measurements are cooling curves and a form of sample controlled thermal analysis whereby the cooling rate of the sample is dependent on the cup material (usually bonded sand) and sample volume which is normally a constant due to the use of standard sized sample cups.

Advanced techniques use differential curves to locate endothermic inflection points such as gas holes, and shrinkage, or exothermic phases such as carbides, beta crystals, inter crystalline copper, magnesium silicide, iron phosphide's and other phases as they solidify. Detection limits seem to be around 0.01% to 0.03% of volume.

In addition, integration of the area between the zero curve and the first derivative is a measure of the specific heat of that part of the solidification which can lead to rough estimates of the percent volume of a phase. (Something has to be either known or assumed about the specific heat of the phase versus the overall specific heat.) In spite of this limitation, this method is better than estimates from two dimensional micro analysis and a lot faster than chemical dissolution.

Foods

Most foods are subjected to variations in their temperature during production, transport, storage, preparation and consumption, e.g., pasteurization, sterilization, evaporation, cooking, freezing, chilling, etc. Temperature changes cause alterations in the physical and chemical properties of food components which influence the overall properties of the final product, e.g., taste, appearance, texture and stability. Chemical reactions such as hydrolysis, oxidation or reduction may be promoted, or physical changes, such as evaporation, melting, crystallization, aggregation or gelation may occur. A better understanding of the influence of temperature on the properties of foods enables food manufacturers to optimize processing conditions and improve product quality. It is therefore important for food scientists to have analytical techniques to monitor the changes that occur in foods when their temperature varies. These techniques are often grouped under the general heading of thermal analysis. In principle, most analytical techniques can be used, or easily adapted, to monitor the temperature-dependent properties of foods, e.g., spectroscopic (NMR, UV-visible, IR spectroscopy, fluorescence), scattering (light, X-rays, neutrons), physical (mass, density, rheology, heat capacity) etc. Nevertheless, at present the term thermal analysis is usually reserved for a narrow range of techniques that measure changes in the physical properties of foods with temperature (TG/DTG, DTA,DSC and Transition temperature).

Printed Circuit Boards (PCB)

Power dissipation is an important issue in present-day PCB design. Power dissipation will result in temperature difference and pose a thermal problem to a chip. In addition to the issue of reliability, excess heat will also negatively affect electrical performance and safety. The working temperature of an IC should therefore be kept below the maximum allowable limit of the worst case. In general the temperatures of junction and ambient are 125 °C and 55 °C respectively. The ever-shrinking chip size causes the heat to concentrate within a small area and leads to high power density. Furthermore, denser transistors gathering in a monolithic chip and higher operating frequency cause a worsening of the power dissipation. Removing the heat effectively becomes the critical issue to be resolved.

Differential Thermal Analysis

In case of DTA reference and sample material is heated at the same rate under controlled conditions and the difference of temperature between reference and sample material is continuously measured against time. This difference in temperatures is plotted as a function of temperature or time and called DTA curves or thermo gram. If temperature difference is zero between reference and sample material then sample doesn't undergo any physical or chemical change, and if there is temperature difference between sample and reference material then physical or chemical change takes place in a sample. These changes result in heat being absorbed (endothermic process) or evolved (exothermic process). Endothermic changes include vaporization, phase changes such as melting, sublimation, transition between two different crystal structures, decomposition and so on; whereas exothermic changes include crystallization, chemisorptions, oxidation reduction and so on. Thus any change in state can be detected by measuring the temperature difference. By convention, endothermic response is represented by downward peaks whereas exothermic response is shown by upward peaks. Curve obtained from DTA can be used for identification purposes as a fingerprint of material. As an example, DTA can be used to study the point when structural resemblance of different forms of clay complicates the interpretation of diffraction patterns. Area under DTA peaks gives enthalpy change of the sample. Furthermore, DTA and TGA are complimentary techniques.

Types of DTA

On the basis of temperature sensing system DTA are of two types:

- Heat flux DTA: In case of heat flux DTA thermocouple is placed outside the sample and reference material.
- Classic DTA: In case of classic DTA thermocouple is immersed into the sample and reference material.

DTA Experimental Factors

Care should be taken while selecting the experimental factors. For example powder decomposition reaction is affected by the specimen environment, size, surface to volume ratio and composition. Although solid state phase changes may not be impacted by these variables. Usually, the experiments involve analysing powder samples so that results do not represent the bulk samples, wherein strain energy builds up to control the transformations. Another factor influencing the decomposition reaction is the packaging of the powders, which leads to large difference in similar samples. Some samples may evolve large amount of heat and may cause saturation of the response capability of measurement systems. To avoid this situation, the sample can be diluted with inert materials. To measure temperature of phase transformations, the maximum temperature should not be varied with sample size. The weight of the sample and the rate of heating do not affect the shape of peaks in DTA. The effect of reducing heating rate is similar to the effect of decreasing the weight of the sample, and both result in sharper peaks with enhanced resolution. However, this is advantageous only in the case when signal to noise ratio is not affected. Studies involving examination of decomposition reactions can benefit from the effects of heating rate on the shape of the peak as well as its disposition. Nonetheless, kinetic studies require minimization of thermal gradients which can be achieved by decreasing either sample size or heating rate.

Instrumentation

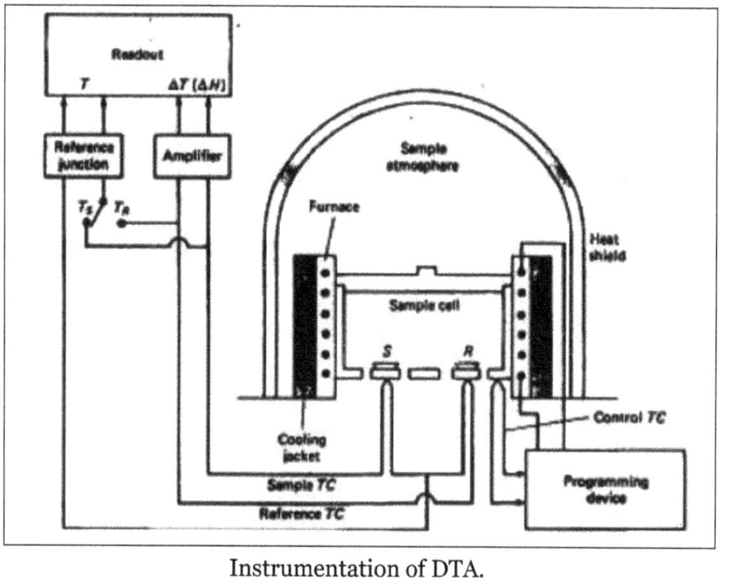

Instrumentation of DTA.

- Sample Holder: Sample and reference crucible are generally metallic (al, pt) or ceramic(silica) and may or may not have a lid, for good results area of contacts b/w sample and crucible is maximized. Typically 1-10 mg of sample is required for analytical applications.

- Furnace: Reference and sample should be thermally matched and symmetrically arranged with the furnace so that both of them are identically cooled or heated, metal block around the wall acts as a heat sink and by using internal heater temperature of the heat sink is slowly increased sink in turn heat the sample and reference material.

- Sensors and recording system: Pair of matched thermocouple is used; one pair is in contact with the sample while the other pair is in contact with the reference. The output of the differential thermocouple ts-tr is sent to the data acquisition system after amplification. Operating temperature for DTA instruments is generally from room temperature to around 1600 °C. Liquid nitrogen cooling accessories is needed for very low sub ambient temperature. Figure shows instrumentation of DTA.

Interpretation and Presentation of Data

A typical DTA plot consists of several linear portions displaced from abscissa due to: (a) the differences in the heat capacity and thermal conductivity of the reference and test sample; (b)physical or chemical changes taking place in the samples result in either absorption or evolution of heat, which is seen as peaks in DTA plots. It is difficult to measure the transition temperatures from DTA plots. This can be understood as follows: In principle, the onset of a DTA peak signifies the start temperature. However, depending upon the relative position of thermocouple with the reference, test sample, or the DTA block, there might be temperature lag. This can be avoided by calibrating the equipment with materials whose melting point is precisely known. The enthalpy change is related to the peak area A, or the area enclosed between the peak and interpolated baseline. If differential thermocouple is in thermal contact and not in physical with the reference and test material then A can be found using:

$$A = \frac{mq}{gK}$$

Here, m is mass of sample, q is enthalpy change per unit mass, g is shape factor and K is thermal conductivity of the test specimen. In case of porous, dense or heaped specimen, the thermal conductivity of the surroundings of DTA container can be altered due to the presence of gas in these pores. The situation further worsens if these gases evolve from the test specimen, leading to a thermal conductivity of the DTA–cell environment which is different from the one used during calibration process. To calibrate the DTA apparatus for enthalpy measurements, area under the peaks of the standard samples is measured over the specified temperature range. A minimum of two samples are required for calibration and both heating and cooling experiments are conducted. The heat capacity at constant pressure (C_p) can be measured as:

$$C_P = K\frac{T_2 - T_1}{mH}$$

Here T_1 and T_2 represent the temperature differences obtained by running the DTA apparatus without the test specimen and with test specimen, respectively. H is the rate of heating and the K is a constant which is measured by calibration against standard materials.

DTA Thermogram

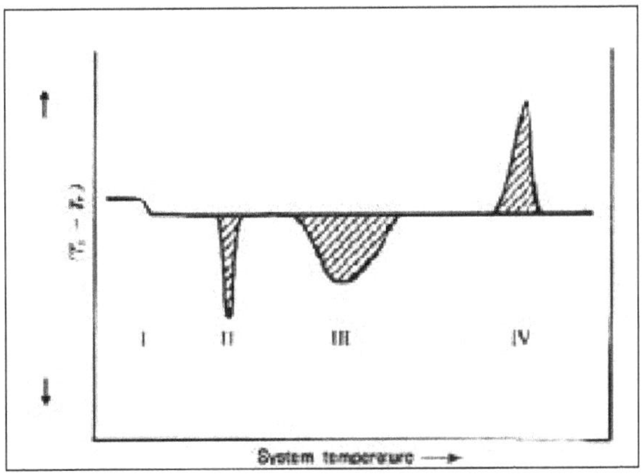

DTA Thermogram.

It is a plot of temperature difference versus temperature as shown in Figure. Four transitions detect by DTA are as follows:

- Second order transition in which change in horizontal line is detected (e.g. glass).
- Narrow endothermic curve due to the melting process.
- Broad endothermic curve due to the exothermic process.
- Exothermic curve due to the crystalline phase changes.

Factors Affecting DTA Curves

Since DTA is a dynamic technique, a large number of factors can affect the resulting experimental curves. If the DTA curve is used for quantitative purposes, the shape, position, the area enclosed by the curve are of great interest.

For specific heat measurements the baseline deviations become important and such conditions as particle size, system symmetry, sample packing must be taken into account if accurate results are to be obtained. As with the technique of thermogravimetry, the DTA curve is dependent on two general categories of variables: instruments factor and sample characteristics.

- Instrumental parameters: It includes furnace atmosphere, size and shape of furnace, sample holder materials, sample holder geometry, heating rate, and location of thermocouple in sample chamber, speed and response of recording device.
- Characteristics of sample: It includes particle size of sample, amount of sample, packing density, swelling or shrinkage characteristic of sample, degree of crystallinity, presence of diluents, thermal conductivity and heat capacity.

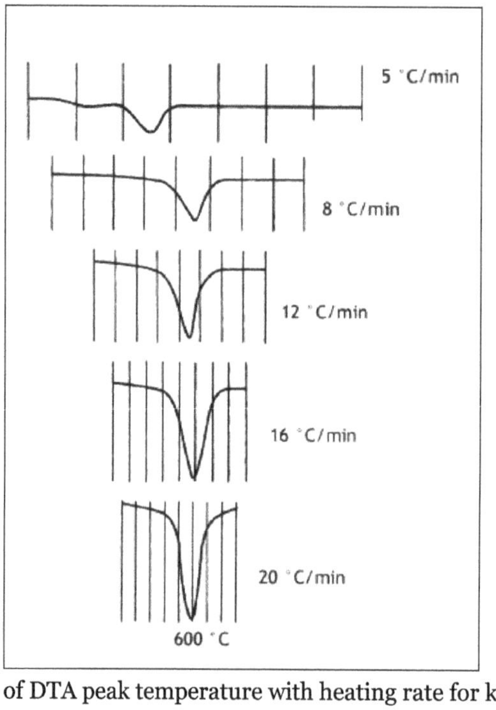

Variation of DTA peak temperature with heating rate for kaolinite.

The effect of furnace atmosphere is similar to that discussed in the thermogravimetry section and is significant for equilibrium reactions. An increase of the heating rate would cause the spreading of the DTA curve. Since the return of the signal to the baseline is a time function, this will happen at a higher actual temperature with more rapid heating is shown in fig.

It is also found that increasing the heating rate will decrease the resolution between two adjacents peaks. The amount of the sample and the heating rate are critical in the dta analysis as well.

A Small Amount of Sample

- Yields maximum resolution of the peaks.
- Yields best quantitative results.
- Yields more regular peak shapes.
- Permits best thermal contacts with the sample holder.
- Allow quick removal of volatiles.
- Minimizes thermal gradients within the sample.
- Permits the use of higher heating rates.

A large Amount of Sample

- Allows detection of small thermal effects.
- Provides more precise qualitative results.
- Provides greater quantities of volatiles for evolved gas analysis.
- The heating rate is normally in the 5-10 C range.

Quantitative Aspects of DTA

Determination of the heat of transformation (reaction), or the mass of the reactive components of the sample, is made from the curve peak area:

$$A = \int_{t1}^{t2} \Delta T \, dt = \frac{m\Delta H}{gK}$$

Where,

- A is the curve peak area.
- Dh is the molar enthalpy change.
- M is the mass of the reactive component.
- K is the thermal conductivity of the sample.

- G is a constant related to the geometry of the sample holder.
- In practice a simplified expression is used: A = mΔH K.
- A is the curve peak area.
- K is the callibration coefficient.

The callibration coefficient is determined by the use of compounds having known heat of transformation, most standard used involve the heat of fusion, dhf or that of a solid (1) solid (2) transition. The calibration coefficient, k is temperature dependent as shown in figure.

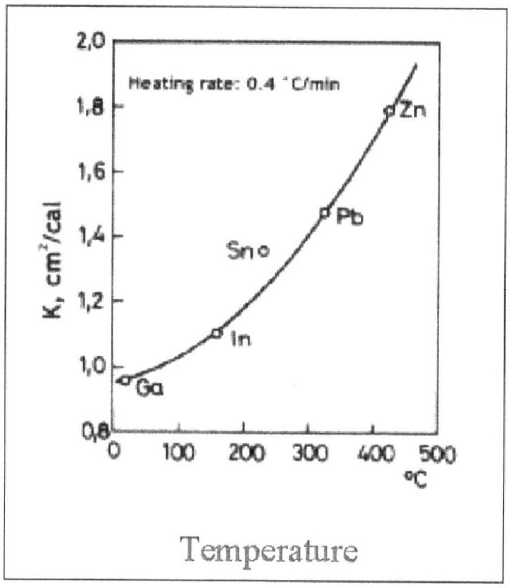

Calibration constant vs Temperature.

Temperature Dependence of Calibration Constant k

This means that the calorimetric sensitivity of the dta apparatus decreases with temperature, i.e. more heat is required per unit area. The peak area can be converted into mass or enthalpy change only if the value of k at the transformation temperature is known, precise determination of the curve peak area is often a problem, since the heat capacity and the thermal conductivity of the sample changes in the course of transformation, as a result, the shifting baseline in the area determination is assumed to be linear (which is not true), or an equivalent peak area is determined which is also an approximation.

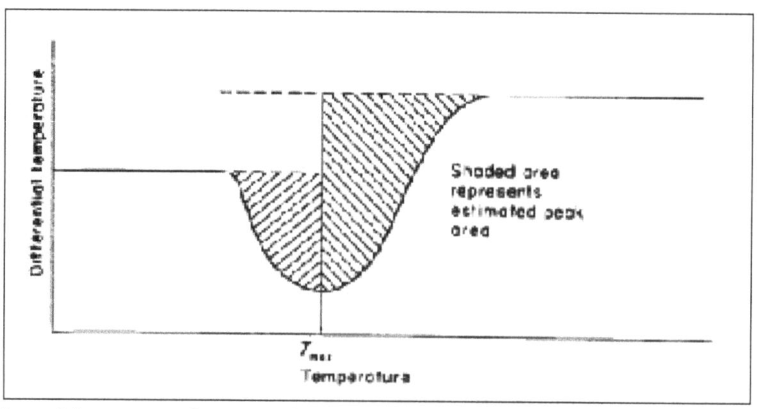

Calculation of the DTA peak area in the case of shifted baselines(change in heat capacity).

Applications of DTA Differential Thermal Analysis

- Analysis of characteristic decomposition patterns.
- Studies of degradation mechanisms and reaction kinetics.
- Determination of organic content and inorganic content in a sample.
- Study of materials, chemical analysis.
- Any change which is associated with enthalpy change.

Differential Scanning Calorimetry

Differential scanning calorimetry (DSC) measures temperatures and heat flows associated with thermal transitions in a material. In this thermo analytical technique, the difference in the amount of heat required to increase the temperatures of a sample and a reference are measured as a function of temperature. Both the sample and the reference are maintained at nearly the same temperature throughout the experiment. Generally, the temperature program for a DSC analysis is designed such that the sample holder temperature increases linearly as a function of time. Only a few milligrams of material are required to run the analysis.

Applications

Common usages of DSC include investigation, selection, comparison, and end-use performance evaluation of materials in research, quality control, and production applications. DSC is commonly used to measure a variety of properties in both organic and inorganic materials, from metals and simple compounds to polymers and pharmaceuticals. The properties measured include:

- Glass transitions,
- Phase changes,
- Melting,
- Crystallization,
- Product stability,
- Cure/cure kinetics,
- Oxidative stability,
- Heat capacity and heat of fusion measurements.

Principle of Operation

When a sample undergoes a physical transformation, such as a phase transition, more or less heat will need to flow to it than to the reference (typically an empty sample pan) to maintain both at the same temperature. Whether more or less heat must flow to the sample depends on whether the process is exothermic or endothermic.

For example, as a solid sample melts to a liquid, the sample will require more heat than the reference to increase its temperature at the same rate as the reference due to the absorption of heat by the sample as it undergoes the endothermic phase transition from solid to liquid. Likewise, as a sample undergoes exothermic processes (such as crystallization), the sample requires less heat than the reference to raise the sample temperature at the same rate as the reference. By observing the difference in heat flow between the sample and reference, differential scanning calorimeters can measure the amount of heat absorbed or released during such transitions. DSC may also be used to observe more-subtle phase changes, such as glass transitions.

Types of Differential Scanning Calorimetry

Various types of DSC are:

1. Power Compensated DSC: In case of power compensated DSC, both reference and sample are heated separately under different furnace. Difference of temperature between reference and sample is maintained at zero i.e. both are at the same temperature and the difference of energy between reference and sample is recorded as a function of sample temperature. Two independent heating unit are employed for power compensation DSC. To rapidly heat, cool or maintaining equilibrium, small heating unit is required. The heating unit needs to be integrated with a relatively larger temperature controlled heat sink. For continuously monitoring the temperature of the material in reference and sample holders, platinum resistance thermometer is used. The increasing the temperature, suitable power is applied to the reference and sample heaters. This power required to maintain equal temperature of reference and sample is recorded by the power compensated DSC as a function of temperature. Power compensated DSC has lower sensitivities than the heat flux DSC, however power compensated DSC are faster than their heat flux counterparts. Thus, power compensated DSC are often used in studying kinetics wherein very fast equilibrium to a desired temperature is reached. These instruments also have higher resolution than the heat flux DSC. The calibration constant in power compensated DSC, which converts peak area into joules, is an electrical conversion factor. Basic diagram of power compensated DSC is shown in Figure.

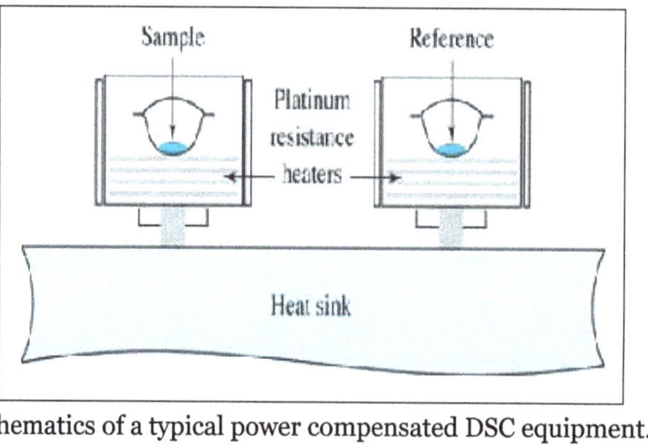

Schematics of a typical power compensated DSC equipment.

2. Heat Compensated DSC: Here, both the reference and sample are heated under the same furnace. The difference between the temperatures of the reference and sample is recorded and converted into a power difference and plotted against time or temperature. The main cell assembly of

DSC is enclosed within a heated silver block cylinder, from which heat is transmitted onto the specimen by a constantan disc connected to the cylinder. The reference and sample pan are kept on two raised platforms in the system. The underside of both the platforms is connected to the constantan disc via connecting wires and chromel (alloy of nickel and chromium) disks, thereby forming a thermocouple which is used to measure the temperature difference between the two platforms. In addition to this, the individual temperatures of the two platforms can be measured by the junction of alumel wires linking the platforms to the chromel disks. Another separate thermocouple is integrated with the silver block which acts as a temperature controller for programmed heating cycles. Inert gas can be introduced within the cell at constant flow rate of ~40nm/min. Unlike the power compensated DSC, the calibration constant (which converts peak area into joule) in heta flux DSC is a thermal factor.

Total heat flow dH/dt in heat flux DSC is:

$$dH/dt = Cp\, dT/dt + f(T, t)$$

Where, H is enthalpy in J/mol

f (T, t) is kinetic response of the specimen in J/mol

Cp = specific heat capacity in $JK^{-1}\, mol^{-1}$.

Therefore, total heat flow is equal to the sum of the two terms; where one term is related to the kinetic response and the other is related to the heat capacity. Figure shows heat compensated DSC.

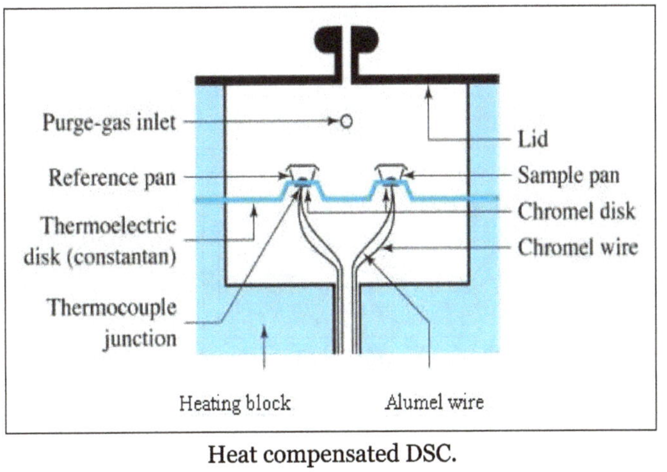

Heat compensated DSC.

3. Modulated DSC: In case of modulated DSC, the cell and heating arrangement are analogous to those used in heat flux DSC. This is a relatively recent technique and was proposed in 1993. A separation of overlapping events in the DSC scans is the main advantage of this technique. In modulated DSC, a linear heating ramp usually is overlapped with a sinusoidal function described by an amplitude and frequency to create a sine wave shaped temperature which varies with time. Fourier transformation can be used to break the DSC signal in two parts: reversible and reflecting events, as:

$$T = T_0 = bt + B \sin(wt)\ dq/dt = C\,[b + Bw \cos(wt)] + f(t, T) + K \sin(wt)$$

Here, T represents the temperature, t is time, C is specific heat, f (t, T) is the average underling kinetic function obtained by subtracting the effect of sine wave modulation, and w is frequency:

[b+Bw cos (wt)] = is the calculated quantity dT/dt or Curve reversing.

K= kinetic response amplitude of the sine wave modulation.

MDSC technique is a valued extension of conventional DSC. Applications of this DSC includes study of the energy of relaxation and precise determination of glass transition temperature.

4. Hyper DSC: New type of power compensating DSC gives high resolution for melting analysis and detection of glass transition temperature (Tg) in medications or crystallisation of metals. This DSC can perform accurate measurements of heat flow with very high linear controlled rates (till 500 K/min) mainly during cooling, i.e., at relatively higher rates than conventional power compensated DSC which works below 10 K per minute. These devices benefit by avoiding unwanted transformation like solid – solid transformation, offer greater sensitivity at higher rates, etc. Hyper DSC is extensively used in pharmaceutical industries at varying temperatures where very high heating rates are required to eliminate various undesirable reactions etc.

5. Pressure DSC: This technique is used to characterize samples at specific environment conditions upto 10MPa as a function of time or temperature. It is often employed in evaluation of catalysts, studying pressure sensitive reactions, and resolution of overlapping transitions.

Instrumentation

Figure shows the block diagram of DSC instrument.

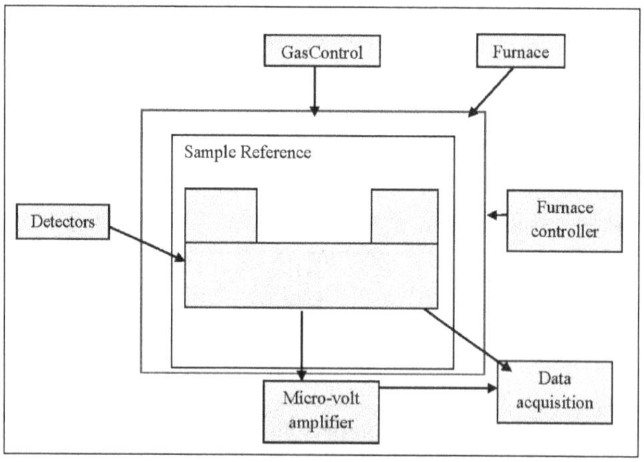

Block Diagram of DSC.

- Detector: In case of detector thermocouple is used to measure the temperature change while the sample heated under controlled environment. For the case of low temperature copper-constantan or cromel alumel thermocouple is used and for the case of corrosive atmosphere or high temperature pt/13%rh thermocouple is used.
- Sample Container: Sample container is designed in such a way that it can be used under high pressure and thermal environment. For high temperature case platinum or ceramic is used for sample holder and for low temperature case aluminium is used.

- Reference material: Generally reference material is made up of calcined alumina Al_2O_3 or carborundum, silicon carbide.
- Temperature controller: Temperature controller is used to maintain the temperature of the reference and sample at the desired value.
- Furnace: One block of furnace is required in case of heat compensated DSC and two block of furnace is required in case of power compensated DSC.

DSC Thermogram

DSC curve is employed to determine the enthalpy of transitions by integrating the peak area for given transitions. The transition enthalpy is found as follows:

$$\Delta H = KA$$

Here, K is the calorimetric constant,

A is the area under the peak,

ΔH is the enthalpy of transition.

The area of a peak scales directly with the heat absorbed (endothermic) or evolved (exothermic) during a reaction, and the peak's height varies directly with the reaction rate. Transitions involve glass transition temperature, melting temperature, crystallization temperature. DSC curve also used to calculate the specific heat capacity of the material using equation:

- Specific heat capacity (Cp) = Absolute value of heat flow/Heating rate.
- Where Absolute value of heat flow = Heat supplied per unit time.
- And Heating rate = Temperature increase per unit time.
- Therefore Cp = $q/t / \Delta T/t$.

  Cp = $q/\Delta T$.

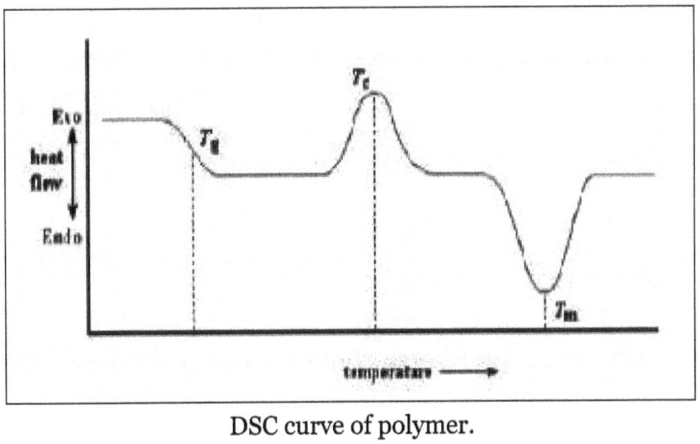

DSC curve of polymer.

Let's consider the case of polymer, initially on heating it absorbs heat and curve look like flat after absorbing certain amount of heat its temperature starts to rise and its heat capacity increase on further heating plot will go downward and polymer reach to its glass transition(second order as there is no latent heat involved) temperature with further increase it comes to its crystallization(-first order latent heat involve) temperature and give off certain amount of heat and reaches to the state of ordered arrangement on further heating it absorbs heat and reaches to its melting(first order) temperature. Figure shows DSC curve.

Factors Affecting DSC Curve

- Instrumental factors: These include parameters such as – Furnace heating rate, furnace atmosphere, recording speed, geometry of sample holders, location of sensors, sensitivity of the recording system and material of the sample container.
- Sample characteristics: Parameters including type and weight of sample, particle size, solubility of evolved gases in the sample, heat of reaction and thermal conductivity.

Applications of DSC

DSC can be used in studying heat of melting, crystallization, percentage of crytallinity, heat capacities, thermal stabilities, purities, oxidative stabilities, food science, drug analysis, and liquid crystals.

Spectroscopy

Spectroscopy involves investigating the interaction of electromagnetic field with matter. Historically, it originated as the study of wavelength based dispersion of visible light by a prism. The idea was expanded to include all the interactions involving radiative energy variation with wavelength or frequency. Spectroscopic data is represented as an emission spectrum of wavelength or frequency dependent response. Spectroscopy can be broadly classified into two categories – (a) techniques based on energy transfer between photon and sample, and (b) reflections, refraction, diffraction, dispersion, or scattering from the sample altering the amplitude, phase angle, polarization, or direction of propagation of the electromagnetic radiation.

Over the years, spectroscopy has evolved as a potential tool for experiments and analyses conducted in research laboratories and industries. This technique is essentially considered by analysts as an apparent solution. The objective should also be to use these spectroscopic techniques in control and industrial laboratories and to develop fully recognised spectroscopic techniques.

Principles and Instrumentation for UV-Vis-IR

Ultraviolet (UV) spectroscopy is an important physical tool which exploits light in ultraviolet, visible, and near infrared range of electromagnetic spectrum. Beer-Lambert law establishesa linear relationship between absorbance, concentration of absorbers (or absorbing species) in the solution and the path length. Therefore, UV-Vis spectroscopy can be employed for determining the concentration of the absorbing species, for a fixed path length. This is a very simple, versatile, fast, accurate and cost-effective technique. Instrument employed for ultraviolet–visible (or UV-Vis) spectroscopy is called UV–Vis–NIR Spectrophotometer. This can be used to analyse liquids, gases and solids by using radiative energy corresponding to far and near ultraviolet (UV), visible (Vis) and near infrared (NIR) regions of electromagnetic spectrum. Consequently, predetermined wavelengths in these regions have been defined as: UV: 300 - 400 nm; Vis: 400 - 765 nm; and NIR: 765 - 3200 nm.

1. Principle: A light beam is passed through an object and wavelength of the light reaching the detector is measured. The measured wavelength provides important information about chemical structure and number of molecules (present in intensity of the measured signal). Thus, both quantitative and qualitative information can be gathered. Information may be obtained as transmittance, absorbance or reflectance of radiation in 160 to 3500 nm wavelength range. The absorption of incident energy promotes electrons to excited states or the anti-bonding orbitals. For this transfer to occur, photon energy must match the energy needed by electron to be promoted to next higher energy state. This process forms the basic operating principle of absorption spectroscopy. Potentially, there may be three types of ground state orbitals involved:

- σ (bonding) molecular orbital.
- π (bonding) molecular orbital.
- n (non-bonding) atomic orbital.

Besides, the anti-bonding orbitals are:

- σ* (sigma star) orbital
- π* (pi star) orbital

A transition involving excitation of an electron from s bonding orbital to σ anti-bonding orbital is called σ to σ* transition. Likewise, π to π* represents the excitation of an electron of a lone pair (nonbonding electron pair) to an antibonding π orbital. Electronic transitions occurring due to absorption of UV and visible light are:

σ to σ*;
n to σ*;
n to π*;
π to π*.

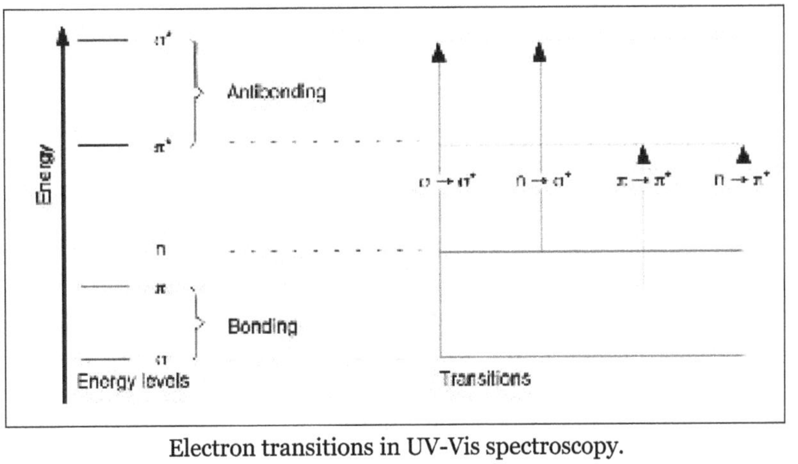

Electron transitions in UV-Vis spectroscopy.

The transitions s to σ* and n to σ* involve higher energies and thus usually occur in far UV region or weakly in 180 to 240nm region. Thus, saturated groups do not show strong absorption in UV region. Molecules with unsaturated centres undergo n to π* and π to π* transitions; these transitions involve lesser energies and thus occur at longer wavelengths than transitions to σ* anti-bonding orbitals.

When light having specific wavelength and energy is focused onto the sample, it absorbs some energy of the incident wave. A photo detector measures energy of transmitted light from sample, and registers absorbance of the sample. The absorption or transmission spectrum of the light absorbed or transmitted by the sample against the wavelength is formed. Bouguer–Beer law or the Lambert–Beer rule is basic principle of quantitative analysis, and it establishes that absorbance of a solution scales directly with analyte concentration. For a given wavelength, the absorbance (unit less) A is described as the molar absorptivity of the absorbing species (M^{-1} cm^{-1}), b is path length of sample holder (normally 1 cm), and c is the concentration of the solution (M).

$$A = a.\ b.\ c$$

UV–Vis–NIR spectrometer can monitor absorbance or transmittance in UV – visible wavelength range. The relation between incident light of intensity 'I_o' and transmitted light of intensity 'I' is described as follows.

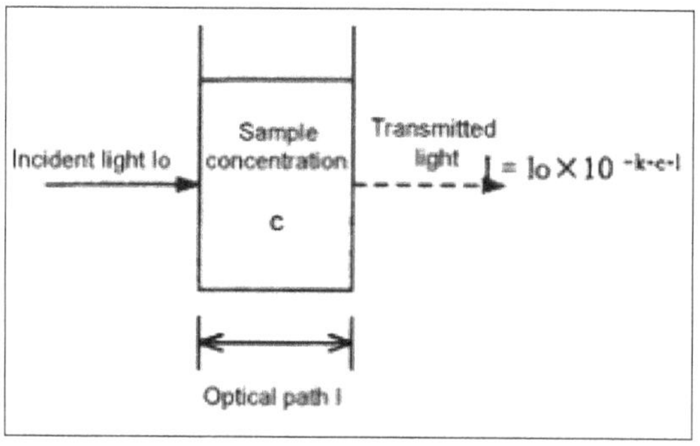

Bouguer–Beer Rule.

Transmittance (T) is given by I/I_o and $(I/I_o)*100$ gives transmission rate (T%). Absorbance (abs) is the inverse of transmittance and given by $\log(1/T) = \log(Io/I)$.

$$T = I/I = 10^{-kcl}$$
$$abs = \log(1/T) = \log(I_o/I) = -kcl$$

Here, k represents constant of proportionality. Transmittance does not depend on sample concentration, whereas absorbance shows proportionality with concentration of sample (Beer's law) and optical path (Bouguer's law). Additionally, when optical path is 1 cm and concentration of targeted material is 1mol/l, k is described as molar absorption coefficient and denoted as 'ε'. Under specific conditions, molar absorption coefficient is typical of the material. Bouguer–Beer rule assumes the absence of any stray, generated, scattered, or reflected light.

The UV-Vis spectrum can by recorded via the following types of absorbance instruments:

- Single bram spectrometer
- Double beam spectrometer
- Simultaneous spectrometer

Light source (mostly tungsten lamps), small holder and detector are common to all the three type of spectrometers. However, a filter may be used, in addition, to select one wavelength at a time. This filter is often termed as the monochromator. Single beam spectrometer (shown in Figure) includes a monochromator between the light source and specimen. The specimen is analysed individually for all wavelengths. Double beam spectrometer uses a single light source, monochromator, a splitter and a series of mirrors, to direct the beam towards the reference and the sample under investigation. Whereas a simultaneous spectrometer uses an array of diodes for simultaneous detection of absorbance at all wavelengths. This is the fastest and most efficient of the three.

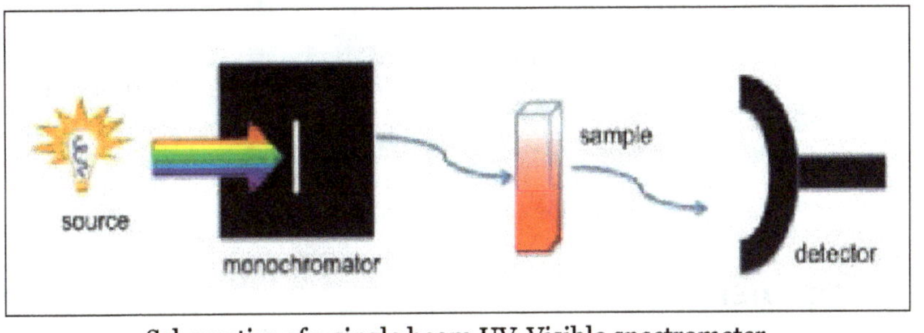

Schematics of a single beam UV-Visible spectrometer.

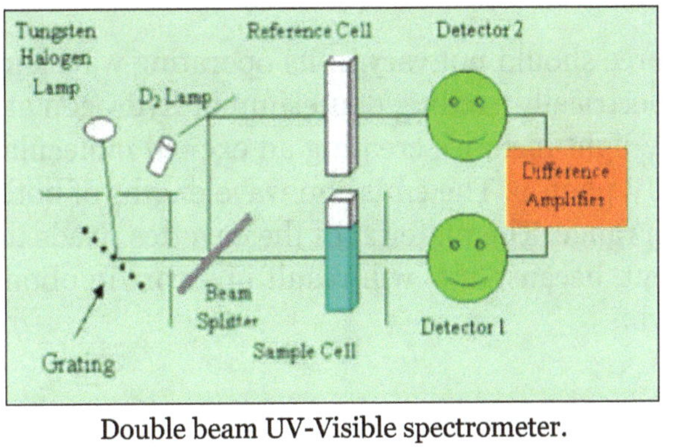

Double beam UV-Visible spectrometer.

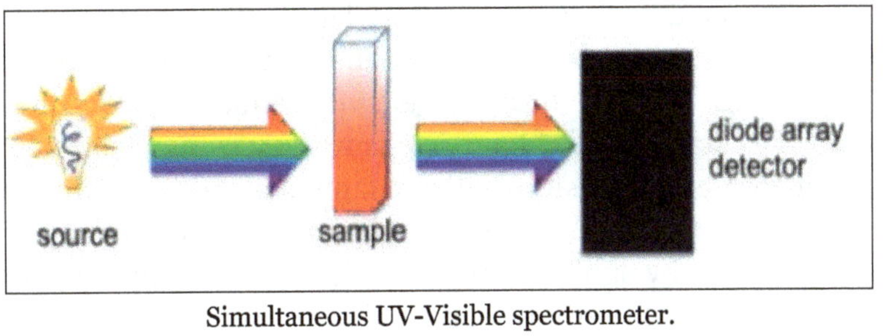

Simultaneous UV-Visible spectrometer.

2. Instrumentation: The basic components of a spectrometer include: light source (UV and visible), monochromator (wavelength selector), sample stage, and detector. A tungsten filament, continuous over UV region is generally used as light source. Detector is usually a photodiode or CCD.

Photodiodes go with monochromators to filter light of a particular wavelength, to be fed to the detector. While monitoring the absorbance in UV spectrum, the visible lamp must be turned off, and vice-versa. Figure includes schematic UV–Vis–NIR Spectrometer.

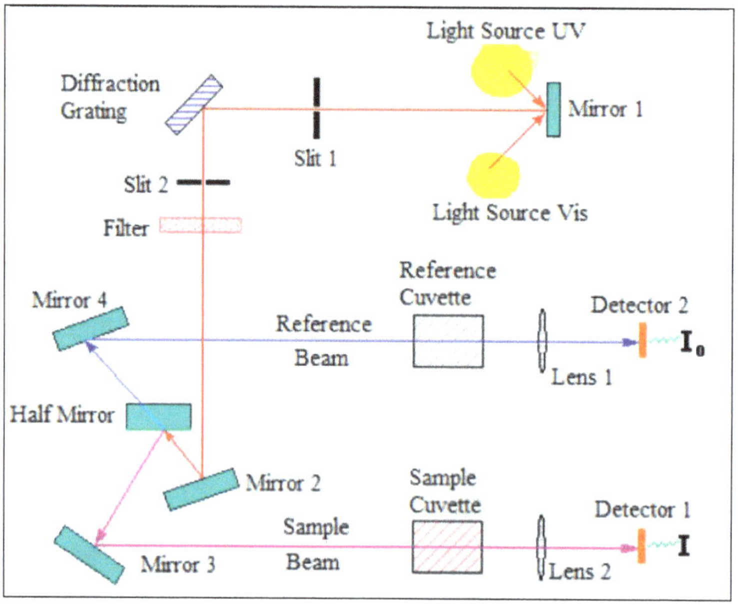

UV–Vis–NIR Spectrometer.

Instrumental Components

UV Source

The power of radiating source should not vary in its operating wavelength range. Continuous UV spectrum is produced by electrically exciting deuterium or hydrogen at low pressures. The mechanism for generation of UV light includes creating an excited molecular species, that breaks into two atomic species and a UV photon. The emission wavelengths of both deuterium and hydrogen lamps are in 160 to 375 nm range. The material of the cuvettes needs to selected such that it does not absorb the light incident, because this will result in errors in obtained absorption spectrum. Thus, quartz is usually used.

Visible Light Source

Tungsten filament lamp is used as visible light source. This lamp can produce light in 350 to 2500 nm wavelength range. In a tungsten filament lamp, energy emitted is proportional to the fourth power of the operating voltage. Thus, in order to get stable emission, a highly stable voltage must be applied to the lamp. The stability of voltage is ensured by using electronic voltage regulators or constant-voltage transformers. Tungsten/halogen lamps include small quantities of iodine embedded within a quartz 'envelope', which also contains the tungsten filament. The iodine reacts with gaseous tungsten, formed by sublimation, and produces a volatile compound WI_2. As WI_2 molecules hit the filament, they decompose, and redeposit tungsten back on the filament. The tungsten/halogen lamps usually have lifetime twice to the conventional tungsten filament lamp. Tungsten/halogen lamps are used in modern spectrophotometers owing to their high efficiency, and their output extends to UV region as well.

Cuvettes

Monochromator source is used; before reaching sample, light is divided in two parts of similar intensity with a half-mirror splitter. One part (or sample beam), travels via the cuvette having the solution of material to be examined in transparent solvent. Second beam, or reference beam, travels via similar cuvette having only solvent. Reference and sample solution containers have to be transparent towards passing beam.

Detectors

Detector detects intensity of light transmitted by cuvettes and sends this data to a meter to record and display the values. Electronic detectors calculate and compare the intensities of light beams. Several UV–Vis spectrophotometers have two detectors – a phototube and a photomultiplier tube, and reference and sample beams are monitored simultaneously. The photo multiplier tube is the extensively used detector in UV-Vis instruments. It includes a photo emissive cathode (electrons are emitted from the cathode when photons strike it), several dynodes (a dynode emits multiple electrons when one electron strikes it) and an anode. The incident photon, after entering the tube, strikes the cathode. The cathode then emits multiple electrons, which are then accelerated towards the first dynode (whose potential is 90V more positive than cathode). The electrons strike the first dynode, leading to the emission of several electrons for each incident electron. These electrons are then accelerated towards the second dynode, to produce more electrons which are accelerated towards dynode three and so on. All the electrons are eventually collected at the anode. By this time, each original photon has produced 106 – 107 electrons. The resulting current is amplified and measured. Photomultipliers are highly sensitive towards UV and visible radiations and have fast response times. However, photomultipliers are used only at low power radiation as high power light may damage them.

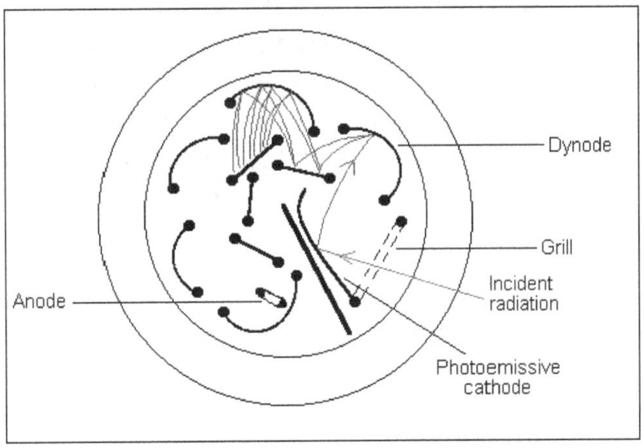

Cross-section of a photomultiplier tube.

The linear photodiode array is an example of a multichannel photon detector. These detectors can simultaneously measure all elements of a beam of dispersed radiation. A linear photodiode array consists of several small silicon photodiodes created on a single silicon chip. The number of photodiodes can vary between 64 to 4096 sensor elements on a chip, however, 1024 photodiodes is most common. For each diode, there is also a storage capacitor and a switch. The individual diodecapacitor circuits can be sequentially scanned.

Charge-Coupled Devices (CCDs) are like diode array detectors, but instead of diodes, they consist of an array of photocapacitors. Reference beam intensity, should suffer little or no absorption, and termed I_o whereas that of sample beam is called I. The spectrophotometer automatically examines all wavelength components in a short time. This technique is good to evaluate the concentration as well as molecular structure or structural changes. It may also be used to examine the vibrational and conformational energy levels alterations before and after an interaction with a substrate, or a molecule.

NMR Spectroscopy

NMR (Nuclear Magnetic Resonance) has developed along with other technical advancements such as highly efficient computers capable of quick Fourier transform, efficient spectrometer control, and stable high-field superconducting magnets. Until past one and a half decade ago, NMR was very rarely used for solid materials. However, the improvements in magnetic angle spinning (MAS) induced spectral resolution has made NMR an important tool for probing the local structure of solid materials. NMR can provide spatial information by applying magnetic field gradients.

Principle

Since nucleus of every element is charged therefore when spins of protons and neutrons within the nuclei do not get paired then total spin of nucleus creates a magnetic dipole along axis of the spin. The magnitude of this dipole is basic property of nucleus and is described as nuclear magnetic moment. Symmetry in charge distribution within nucleus is a function of the internal structure and shape. For spherical distributions, e.g. 1s hydrogen orbital, spin angular momentum number is I = 1/2. Thus, magnetic moment magnitude in a particular direction can have two equal and observable values corresponding to the spin quantum numbers = +1/2 and −1/2. This implies that any nucleus immersed in a magnetic field H along z-direction, may be considered as aligned either with the field $(I_z = -1/2)$ or against it $(I_z = +1/2)$. Analogous to the compass needles in earth's magnetic field, favorable energy state belongs to the alignment along the field. Energy difference (ΔE) between the states scales directly with the applied magnetic field strength (H), or $\Delta E = \gamma\hbar H$, here, $\hbar = h/2\pi$, h = Planck's constant, and γ is the constant of proportionality and depends on the type of nucleus (^{1}H, ^{13}C, ^{15}N, etc.). Similarly, nuclei that possess non-spherical charge distribution like a hydrogen 3d orbital have higher spin numbers, for example ^{10}B, ^{14}N etc.

The application of external magnetic field (B), leads to the aligning of magnetic dipoles of NMR active nuclei in S (S = 2I +1) orientationsor spin states, relative to the field (1). This alignment is limited by quantum mechanics. The corresponding nucleus precess at a frequency, *n*, can be given as:

$$\nu = \frac{\mu\beta_N\beta_o}{hI}$$

Where, μ represents magnetic moment of the nucleus, and β_N = nuclear magneton constant. The magnetogyric ratio (γ) can be expressed as:

$$\gamma = 2\pi\mu\beta_n hI$$

So, Equation $\nu = \frac{\mu\beta_N\beta_o}{hI}$ may simplify to:

$$\nu = \gamma\beta_o / 2\pi$$

The following conclusions can be reached here:

- μ directly controls the sensitivity of NMR experiment, such that the sensitivity increases with a corresponding increase in the value of μ.
- Local field experienced by the nucleus is B_0 (≠B). It is affected by various electronic factors (mainly induction and anisotropy) which control the local density of electrons around the nucleus.
- Inherent frequency of a nucleus is constant, whereas its precessional frequency ν is dependent on B_0.

Different spin states have different energy levels, where RF frequency photons can bridge ΔEs between these levels. Promotions among spin states occur when the incident frequency exactly matches ν ('resonance condition'). Additionally, the excited nuclei can also relax to lower spin states (this is called spin flipping). Excited nuclei can relax via several mechanisms, the most frequent among these are: spin-lattice-relaxation and spin-spin relaxation. Both of which are non-radiative processes. Spin-lattice relaxation involves transfer of energy from excited nucleus to an electromagnetic vector (such as a molecule of polar solvent or intramolecular group experiencing vibrational-rotational processes). On the other hand, spin-spin relaxation involves extra energy to a similar relaxed nucleus. In addition to this, the excess energy may also be re-emitted by the excited nucleus. The effective excitation is governed by the rate of relaxation. Inefficient transfer of energy (e.g., in non-viscous media wherein molecules are randomly oriented) increases the mean half-life of relaxation (≈1 s) and results in sharp spectral lines.

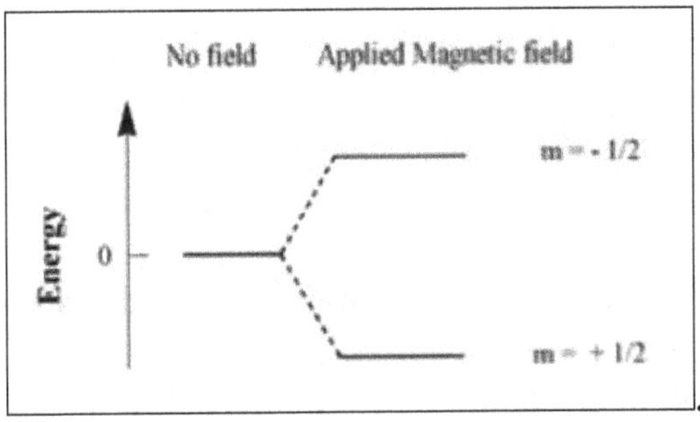

Energy levels for a nucleus with spin quantum number (I) = ½.

Relaxation Processes

Boltzmann distribution describes the thermal equilibrium energy level distributions in materials where the interactions among nuclei are weak. Thermal equilibrium is reached via exchanging energy with the surroundings, or the lattice. The time (T1) required to reach thermal equilibrium is characteristic of the material and is termed as spin-lattice relaxation time. Let us consider the behavior of the macroscopic magnetization M_z which is the vector sum $(\Sigma\mu_z)$ of nuclear dipole components along the polarizing field B_z. Achievement of the equilibrium longitudinal magnetization Mois usually expressed by a simple exponential recovery, depending upon the initial state. Assuming $M_z = 0$ at t = 0 yields:

$$M_z(t) = M_0\left(1 - e^{-t/T_1}\right)$$

To achieve the initial state, a 90° pulse is applied and $M_z(t)$ is experimentally measured from the signal following a second 90° pulse. Therefore, RF pulse sequence $90° - \tau - 90°$ produces a signal, whose amplitude is given by: $V(\tau) = V(0)\left(1 - e^{-\tau/T_1}\right)$ and its measurement as a function of τ, leads to the determination of T_1. Stimulated transitions across the nuclear spin energy levels are required in spin– lattice relaxation. These transitions are the result of fluctuating interactions, which are caused by local magnetic fields at Larmor frequency.

An alternate equilibrium condition for nuclear magnetization requires a zero transverse magnetization (M_{xy}). Transverse magnetization can be produced by applying a coherent electromagnetic radiation at resonant frequency (e.g. a 90°pulse). The maximum magnitude of transverse magnetization is equal to the equilibrium magnitude of the original z component. Various local fields may be generated by either external (such as any inhomogeneity in B_z) or internal (such as fields caused by nearby nuclear dipoles) factors which result in variations in precessional frequencies throughout the sample. This results in loss of transverse coherence after a characteristic relaxation time T_2, which is known as the spin–spin or transverse relaxation time. Typically, magnetization decay in transverse direction is exponential in case of liquids. Measuring M_{xy} with variation in time offers information about the 'local fields', which is important to study microscopic level structure as well as dynamics of materials.

Instrumentation

NMR spectrometer as depicted in Figure applies RF pulse to the sample (under investigation). During the experiment, the sample is placed inside a high magnetic field and the time response of its spin after application of pulse is measured. Magnet and probe are central to the working of the equipment. High magnetic field is critical to the analysis, and the highest commercially available magnetic field is 17.55 T which is usually employed in high-resolution investigations. Standard equipments can generate anywhere between 4.7 to 9.4 T. Magnets are usually prepared from Nb_3Sn or NbTi multi-filament wire based superconducting solenoids, placed within liquid helium are more common. Nonetheless, new and improved materials are required to generate still higher magnetic fields, necessary in nanoscience and nanotechnology investigations. Furthermore, the magnetic field must also be homogeneous. High magnetic field results in improved sensitivity (by enhancing the Boltzmann factor) as well as resolution (by boosting the chemical shift dispersion). But, when line broadening is entirely decided by chemical shift anisotropy, higher magnetic fields adversely affect the powder line widths, even though it reduces the second-order quadrupole effects. Receptivity (R_x) is determined to determine how readily different nuclei can be observed:

$$R_x = \gamma_x C_x I_x (I_x + 1)$$

Here C_x represents the natural abundance of the nucleus.

In case of solids, the RF transmitter should produce short (μs) and intense (~ 1 kW) pulses. Bloch vector model is used in rotating reference frame to model the effect of these pulses on a spin-1/2 nucleus. A pulse of duration T_p applied preciselyat resonance creates a resultant field, orthogonal to B_0 in the rotating frame of reference which further leads to a coherent rotation of the magnetization such that it is tipped by an angle θ_p away from B_0 towards the transverse (xy) direction with:

$$\theta_p = \gamma B_1 T_p$$

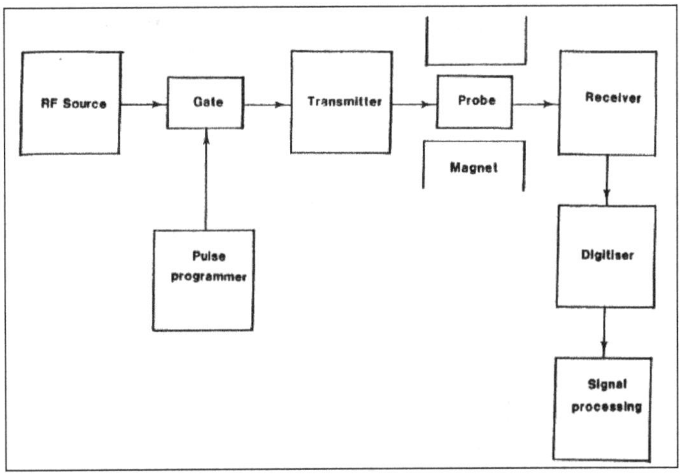

Block diagram showing essential components of a NMR spectrometer.

Following the pulse, the magnetization precesses freely in B_o (at $\omega_o = \gamma B_o$) and may be used for inducing a voltage in some nearby coil. This signal is known as free induction decay (FID). The signal-to-noise ratio (SNR) can be improved by coherent addition of n FIDS. Figure shows the cross- section of magnet used in NMR spectrometers.

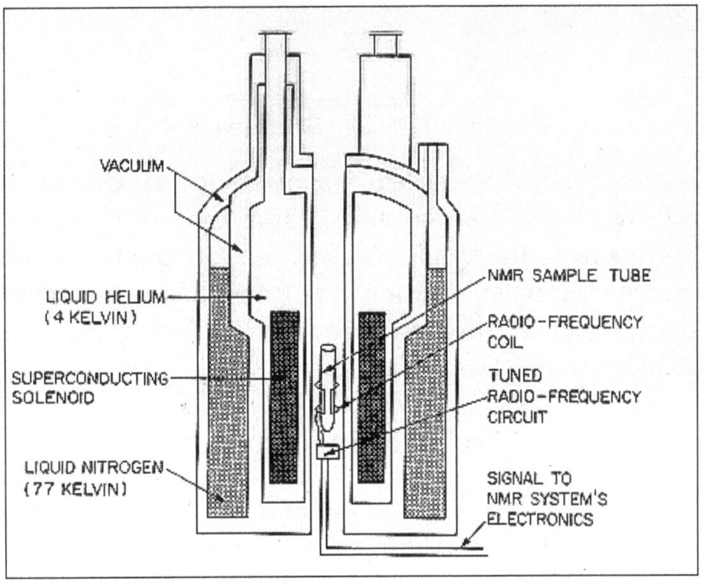

Cross-section of superconducting magnet in an NMR spectrometer.

In a typical NMR investigation, the specimen is positioned in a coil and introduced in a static magnetic field as shown in Figure. Due to the magnetic field, nuclear spins get polarized along z-axis. Besides this field, a transverse magnetic field B1 can also be applied along the x-axis, which can be generated by applying an AC current in coil having Larmor frequency which is given by the following equation:

$$\omega_L \gamma_N B_o$$

By doing so, the resonance frequency of the circuit shown in figure gets changed by varying capacitances of both capacitors C and C'. By matching impedance of power supply (50 Ω) with that of the rest of the circuit, maximum power can be retrieved from the power supply. For appreciating effect of B_1 on the nuclear spins, it is appropriate to describe a rotating frame of reference that rotates about z axis at Larmor frequency.

When AC current in the coil is switched on and off, pulsed B_1 magnetic field is generated along the x axis, this field is the sum of two components, a clockwise and another anti-clockwise rotating fields. The component that remains stationary in the rotating reference frame is significant and is considered further. The spins respond to this pulse by generating a net nuclear magnetization vector to rotate in the direction of the applied field, B_1. Rotation angle is affected by on-time of field, τ, as well as its magnitude B_1.

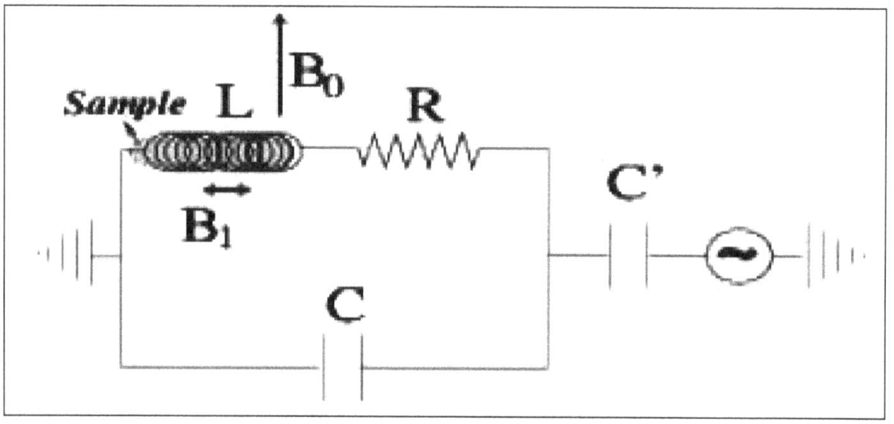

Resonance circuit for the NMR probe.

A $\pi/2$ pulse causes a rotation of 90° in nuclear magnetization about x axis down to the y axis in clockwise direction while in case of the laboratory frame, the equilibrium nuclear magnetization spirals down around the z axis to the xy plane. Thus the rotating frame of reference is useful to describe the behaviour of nuclear magnetization in response to a pulsed magnetic field. Likewise, a π pulse rotates the nuclear magnetization vector by 180°. If the initial nuclear magnetization was along z axis, it is rotated down along the z axis.

Principle and Instrumentation: NQR Spectroscopy

NQR (nuclear quadrupole resonance) spectroscopy is a widely used technique for studying solid-state materials. It is mostly used to determine the effects of pressure and temperature on nuclear quadrupole resonance signal. Following sections describe the instrumentation and working of NQR.

Principle

NQR exploits the interaction between electric field gradient (q) and quadrupole moment (Q), where Q displacement of mean distribution of nuclear charge from the spherical symmetry. The nonspherical nucleus placed in electrical field gradient is shown in Figure. The spinning nucleus may have its axis of spin along the direction of elongation (+Q) or compression (–Q). In this regard, a particular nucleus can have either +Q or –Q spin, but not both. Proton charge is represented by e and q denotes average electric field gradient along z-axis of the principal axis system. Energy level distribution is quantized along q. Quadrupole coupling constant, which defines the strength of interaction of spinning nucleus with q, is determined from the product of quadrupole moment, proton charge, and average electric field gradient, i.e., eQq (or e^2Qq, however the dimensions of q change in this case). To detect NQR signals, the obtained line frequencies are mapped to quadrupole coupling constants, this leads to the direct evaluation of the magnitude of nuclear quadrupole coupling constant (eQq) from line frequencies.

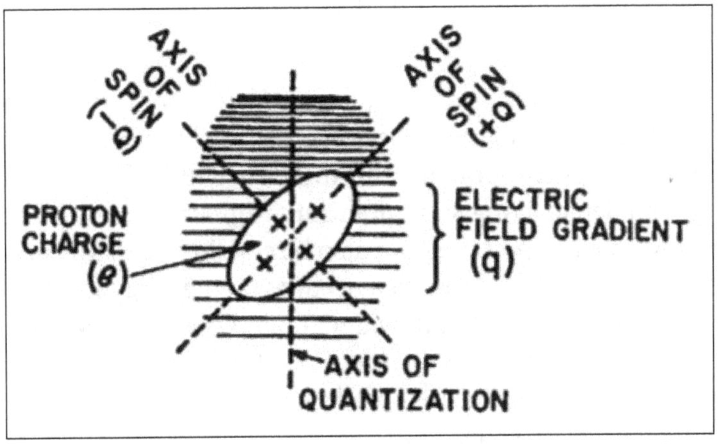

Schematic depiction of non-spherical spinning nucleus (Q) and its interactions with electric field gradient (q).

Among the available elements in periodic table, as many as 70 of them possess at least one isotope with quadrupole moment (Q). The elements exhibiting NQR signals are shown in Figure. As evident from the figure, several low atomic number elements (e.g., lithium, beryllium, boron, hydrogen, nitrogen, etc.) demonstrate NQR activity in all or at least one of their isotopes. Of these elements, nitrogen and chlorine have attracted much interest, where NQR has been used to examine their resonance properties.

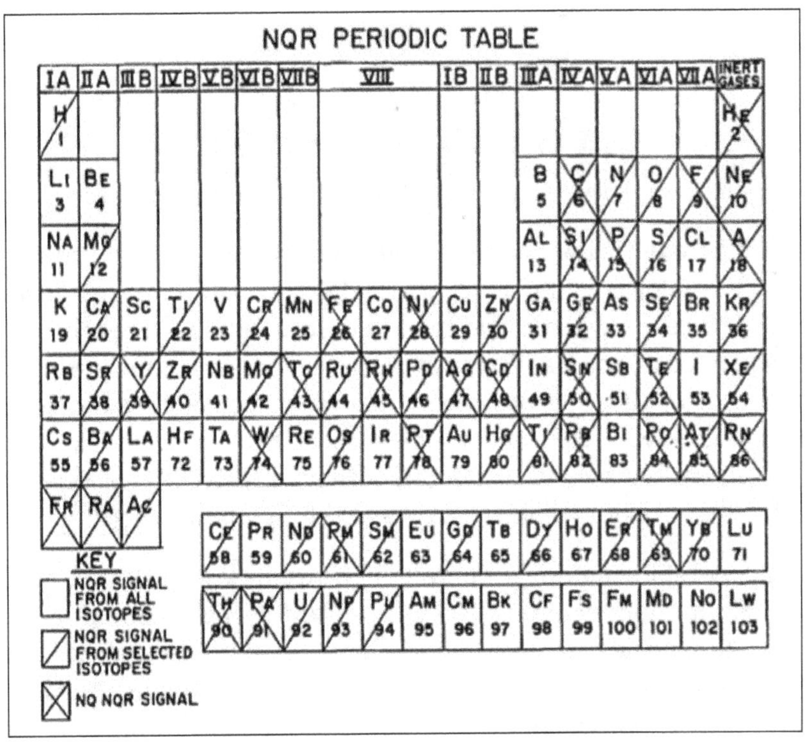

Periodic table showing elements having quadrupole moments.

Nonetheless, NMR study are difficult to perform, for instance, NMR analysis of ^{14}N (main isotope of nitrogen) suffers due to its quadrupole moment, which causes broad lines in NMR spectrum. On the other hand, ^{15}N can be readily used in NMR, as it does not exhibit quadrupole moment. However, it is only ~0.4% in nature, making the achievement of resonance rather difficult to detect. In this case, highly sophisticated computerized techniques such as Fourier transform can be used to investigate the resonance.

Analogous to ^{14}N, the isotopes of chlorine (Cl-35 and Cl-37) both exhibit quadrupole moment, precluding NMR analysis of these isotopes. Nonetheless, resonance signals (for NQR study) are easily detectable in chlorine. Figure highlights this difference in NMR and NQR, wherein the energy levels are drawn for a nucleus having spin quantum number of 3/2. As the isotopes of chlorine also have spin quantum number as 3/2, this can also be considered as the energy level diagram for chlorine in NMR and NQR. In NMR, resonance occurs when energy level split due to a magnetic field (H_0). Thus, resonance is absent in NMR, when applied magnetic field is zero. The application of a non-zero magnetic field splits the energy levels into four. Consequently, a single type of chlorine can undergo several transitions, thereby leading to a broadening of spectral lines in NMR. Additionally, it has short relaxation time.

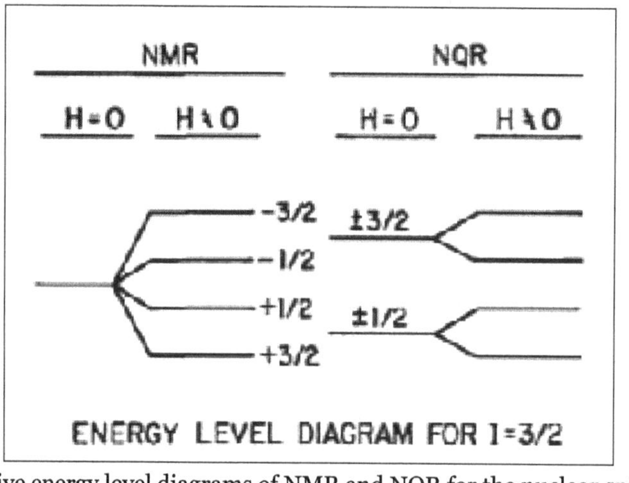

Comparative energy level diagrams of NMR and NQR for the nuclear spin of 3/2.

However, in case of NQR, there are two energy levels for a material even when no field is applied. These energy levels are: $M_j = \pm 1/2$ (lower), and $M_j = \pm 3/2$ (higher), which allow one transition for one type of chlorine in the material. A weak magnetic field splits both these levels, thereby broadening the line and a minimum of two lines appear for a single type of chlorine. For correlating the line frequencies with a particular structure, it is necessary to generate as simple spectrum as possible. Therefore, a magnetic field is usually not applied in NQR experiments unless specific information (e.g., quadrupole coupling constant, asymmetry parameter, etc.) is required. Consequently, NQR is preferred over NMR to observed resonance signals from isotopes having a spin quantum number of 3/2.

NQR is performed in solid state materials whereas high-resolution NMR involves liquid or solution state materials. As a result, the averaging effect as observed in high-resolution NMR is not present in NQR experiments. Thus, nuclear spin system while interacting with electric field gradient is influenced by bond orientations. The dependency over the orientation of bond arises from the contribution of bond electrons tensor components and to the effect of neighbouring ions and/or molecules in the crystal. A three-dimensional Cartesian coordinate system comprises three field gradient tensor component (q_{zz}, q_{yy}, and q_{xx}). Out of which, q_{zz} is the principal component of field gradient tensor, and is in the direction of principal axis of quantization. These components of field gradient tensor are used to define η or the asymmetry parameter, which is equal to $(q_{xx} - q_{yy})/q_{zz}$ and varies between zero and one. η = 0 corresponds to an axial symmetry along z-axis, while η = 1 corresponds to the two-dimensional gradient effect. η is usually described in percentage from 0 to 100%.

The energy level diagram for a spin quantum number of 3/2 in the absence of magnetic field and considering η is included in Figure. η = 0 corresponds to energy levels $E_{\pm 1/2}$(lower) and $E_{\pm 3/2}$(upper), and one transition occurs at v_0. Further, η ≠ 0 also corresponds to two energy levels allowing one transition. In this case, the quadrupole coupling constant (eQq) changes to account for η. Thus, η needs to be determined for calculating quadrupole coupling constant from the resonance frequency for any material.

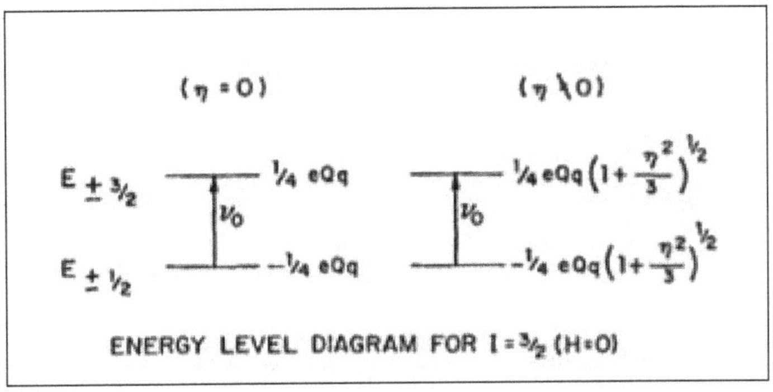

Comparison of the energy level diagram between asymmetry parameters of 0 and greater than 0 for a nuclear spin of 3/2.

For instance, η nearly zero for chlorine, and most of the organic compounds. The situation is different for inorganic compounds because the relaxation time may be substantial to contribute and provide structural information.

Instrumentation

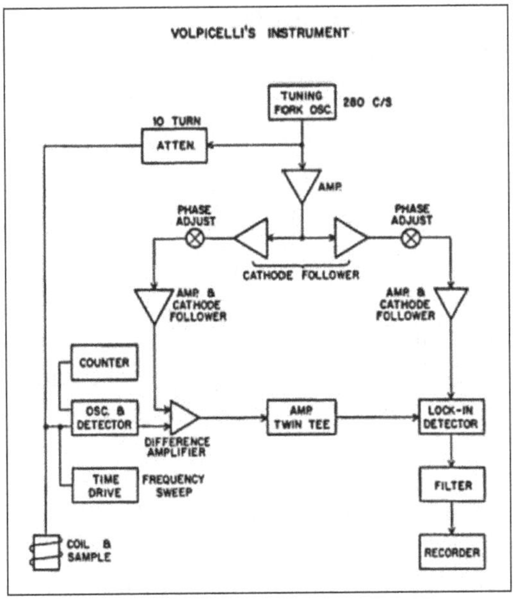

Regenerative oscillator-detector NQR instrument.

Figure shows schematics of an oscillator used to detect NQR signals. This instrument works in 2 to 50 MHz frequency range. The signal from the sample is passed through a differential amplifier that cancels the undesired signal from it. The output of the amplifier drives a lock-in detector which is tuned at the same frequency as the differential amplifier. The signal from the detector is used as the error signal to lock the spectrometer to a quadrupole resonance.

Figure shows the signal from lock-in detector for ^{35}Cl resonance of $KClO_3$, which is similar to that of spectrometer which operates on frequency modulation of magnetic field.

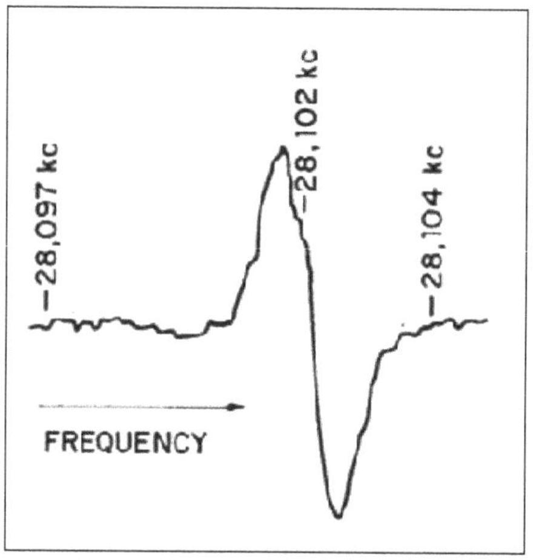

Output of the regenerative oscillator-detector instrument for ^{35}Cl resonance signal of $KClO_3$.

Raman Spectroscopy

Raman spectroscopy has the potential to answer a number of questions pertaining to chemical details of molecular structure what makes this technique definitely proper for material identification. Raman spectroscopy has recently proved progressive benefits and become a powerful research tool which finds assertion in almost all sciences and a wide range of technical and industrial branches. Doubtless advantages as non-destructiveness, contactless measurements, rapidity or no demand for sample preparation makes Raman spectroscopy attractive, convenient and effective. In the light of these facts, the popularity of Raman spectroscopy is still growing in analytical laboratories all over the world. Although the fundamental principle of this method is known almost one hundred years, extensive potentialities which Raman spectroscopy offers are finally able due to a technical advancement and novel engineering solution of the recent years.

LASER spectroscopic research yields essential knowledge necessary for the insight into the world of atoms and molecules dimensions. Especially Raman spectroscopy is a powerful analytic method providing detailed and specific information at a molecular level. In regard of its versatility this method can offer information that can be below possibilities of other spectroscopic methods. From a general point of view an interaction of electromagnetic radiation with a matter, particularly a specific emission and absorption of light, provides essential and valuable acquaintances of chemical agents.

Raman effect was discovered more than eighty years ago. For a long time it remained only on a theoretical level and practically was marginalized because of low sensitivity and difficulties with overcoming fluorescence phenomena. However, renaissance of Raman spectroscopy is coming even in the last two decades hand in hand with technical advancements as are new extremely sensitive detection devices latest developments, efficient filters for filtering Raman scattered light from the Rayleigh scattering and introducing lasers as light sources and also innovative laser technology designs.

The fact, that Raman spectroscopy is very flexible method in majority of sciences and technical branches, has been repeatedly proven during recent years, when problems from different scientific areas have been solved.

A structure and a composition of materials influence their physical utility properties and consequently characteristics of products fabricated from these materials. In many cases strict and specific conditions that must be complied are given. Critical properties on the possibility borders of examination must be attested. Raman spectroscopy is a powerful tool for such verification within a wide range of scientific disciplines. This rapid technique allows quick identification of materials, gives information on the structure, spatial arrangement of molecules and structural analyses. Analogically to the reality that every human being in the world has different – unique – set of fingerprints by which can be identified, also every individual substance has its own unique Raman spectrum characteristic only for the respective substance. Except material identification Raman spectroscopy also enables observation of the structural changes dynamics, mapping of samples and depth scanning.

With regard to Raman molecular structure fingerprint this analytical method is suitable particularly for qualitative analyses, but can be also used for quantification, because the intensity of Raman scattered light is related to the amount of particular substance in a sample. Quantitative studies require often non-linear calibrations and standardization of the Raman signal. When doing quantitative analyses it is crucial to have as stable instrument as is possible, to assure measurement parameters identity as are laser power, time of irradiation, spectral resolution, number of acquisitions or sample orientation and also consider data post-processing. Obviously doing quantitative analyses of liquids and (water) solutions is better and more proper than analyses of solids because of their bigger homogeneity.

Raman and Infrared Spectroscopy

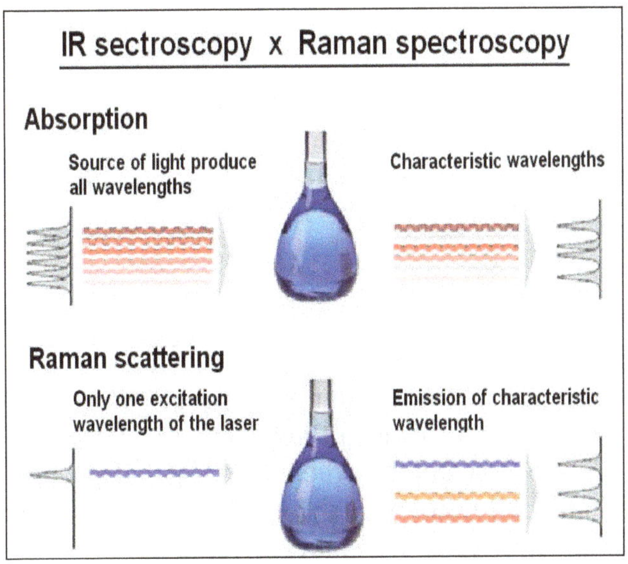

Differences of light-matter interaction of infrared and Raman spectroscopy.

Raman spectroscopy and infrared spectroscopy are adverted as complementary techniques. Both methods provide information about vibrational and rotational transitions in molecule and serve as a good tool for material identification even though principles of these methods are different. However, Raman spectroscopy is not so popular for the chemical analysis of materials as is infrared spectroscopy.

Infrared spectroscopy is an absorption technique dealing with the fact that molecules absorb specific frequencies that define their structure. The output data are absorption spectra. On the other hand Raman spectroscopy provides emission spectra originated during interaction of the monochromatic excitation laser beam with molecules, which emit on for the matter characteristic frequencies. This elemental difference is shown in Figure.

Certain substances have high-quality both Raman and IR features, but for instance centrosymmetric molecules modes give a rise either Raman or infrared spectra but not both. Strong infrared absorptions usually appear as weak Raman signal and vice versa. Infrared spectroscopy has more intensive signal for small polar molecules and single bonded chemical compounds like C-C, C-O, C-H etc. Large absorption bands of water caused by containment of high absorptivity values functional groups disallow infrared measurement of water solutions. On the contrary presence of water in a sample can be determined with ease. Intensive spectra of polyatomic inorganic chemical compounds and majority of organic matters yields Raman spectroscopy. Raman spectrum of water is minimal thereby substances dissolved in water can be identified without problems. A demonstration of infrared and Raman spectrum of styrene-butadiene rubber can be seen in Figure.

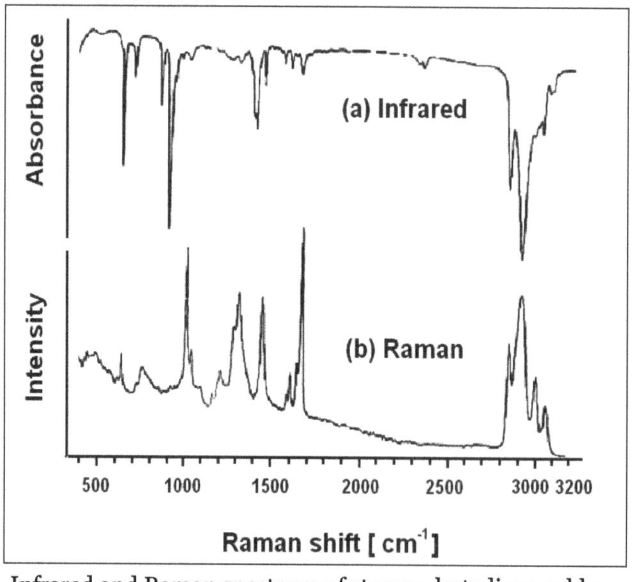

Infrared and Raman spectrum of styrene-butadiene rubber.

From time to time a complication with other quantum process fluorescence, which occurs with a time delay, can appear and disrupt acquiring of Raman spectra. Raman scattering, the fundamental principle of Raman spectroscopy and the whole technique are named in honour of one of its discoverers, the Indian scientist Sir Chandrasekhara Venkata Raman, who searched out the effect together with Kariamanikkam Srinivasa Krishnan in 1928. Sir C. V. Raman was awarded a Nobel Prize in physics in 1930 for the discovery of the effect and his work on the scattering of light.

Sir C. V. Raman and Sir K. S. Krishnan.

As is known, molecules have an ability to absorb or emit photons – an electromagnetic radiation with specific energy and by that change their own energy. According to the quantum theory the energy of a photon E is given by the equation:

$$E = h\nu$$

Or

$$E = hc\bar{\nu},$$

Where h is Planck's constant, ν is its frequency, $\bar{\nu}$ is the wavenumber, thus the reciprocal wavelength. Mentioned specific energy ΔE have to fulfill following quantum condition:

$$\Delta E = \left|E_p - E_q\right| = h\nu = \frac{hc}{\lambda}.$$

E_p and E_q are energies of different quantum states in which a molecule can exists, c is the velocity of light and λ is the wavelength. So the energy of a molecule will be changed by an amount E, if the molecule absorbs or emits the photon.

Raman effect occurs when a researched material is irradiated by an intense monochromatic light. A major part of light beam usually from near infrared, visible or near ultraviolet range is scattered without changes in frequency, no energy is gained or lost, i. e. Rayleigh (elastic) scattering, a part is absorbed and a remained tiny fraction, important for the origin of the spectra, is non-elastically scattered. After interaction of the photon with the molecule, particularly with the electron cloud and the bonds of the molecule, the photon evokes molecule excitation from the ground state to a virtual energy state.

When the molecule relaxes it emits a photon and it returns to a different vibrational or rotational state. The energetic difference, between the ground state and the final state, results in a shift in the emitted photon's wavelength. Energy level diagram illustrates Figure. The schematic principle of the Raman spectroscopy is shown in figure below.

If the molecule absorbs energy, i.e. the final state is more energetic than the initial state then the emitted photon of a lower frequency generates a Stokes line. If the molecule loses energy, the emitted photon of a higher frequency generates an anti-Stokes line. Wavenumber (Raman) shifts carry analytical information on differences between the individual quantum levels and play the key role in substance identification. These Raman shifts depend on the specific molecular geometry of the material and are independent on the incident photon wavelength, i.e. on the excitation wavelength of the laser. There can exist several characteristic shifts for certain material which originates the Raman spectrum.

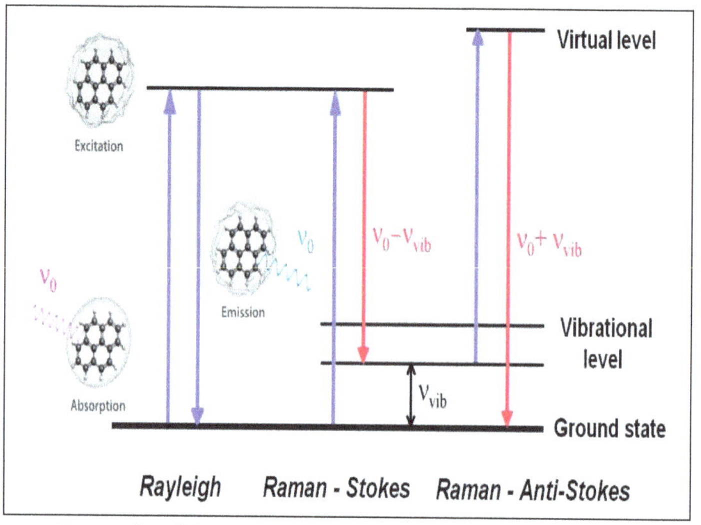

Energy level diagram of Rayleigh and Raman scattering.

Mostly only a more intensive (Stokes) part of the spectrum is measured. Both Stokes and anti-Stokes are approximately symmetrical towards the zero shift of the wavelength that corresponds with the incident laser line wavelength. Distribution of the lines in the spectrum informs about a sort of the bonds in the molecule. Every individual substance has its own unique Raman spectrum characteristic only for the respective substance.

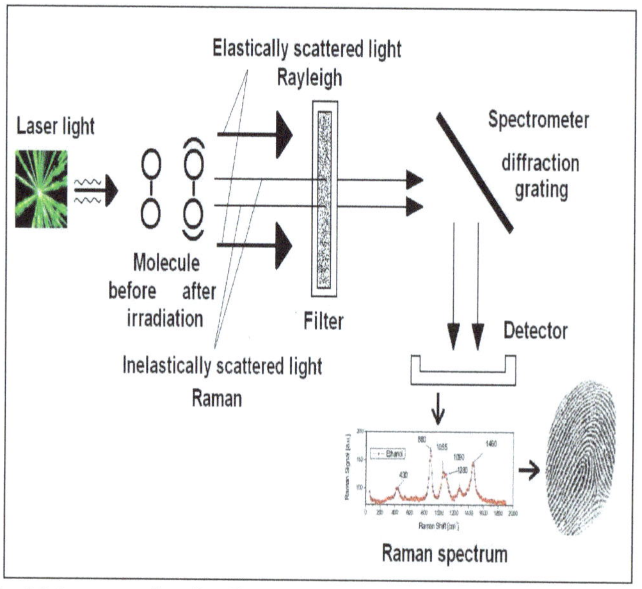

The sample is irradiated with laser, molecule vibrates, filter eliminates intense Rayleigh scattering, the grating disperses the light onto a detector to generate a spectrum, which gives the information about molecule bonding and provides a chemical fingerprint utilizable for identification.

Raman spectrum represents a dependence of intensity of the scattered light (in arbitrary units) on wavelength or on a Raman frequency shift (measured in cm^{-1}). Intensity of the Raman scattering depends on several factors as the excitation wavelength of the used laser, used excitation power, changes in polarizability, the amount of Raman active molecules illuminated by the laser beam and temperature. The intensity of a Raman band is theoretically described by George Placzek.

In spite of the fact, that the particularity of Raman spectroscopy is remarkable, the conversion efficiency of Raman effect is rather poor, since only a scarcity (about 10^{-7}) of the initial photons are non-elastically scattered. Hence the detection of very low concentrated molecules is limited.

In order to enhance sensitivity, to improve intensity, to reach better spatial resolution and other improvements number of variation of Raman spectroscopy has been developed: Surface Enhanced RS, Resonance RS, Transmission RS, Spontaneous RS, Tip-Enhanced RS et al.

Advantages and Disadvantages of Raman Spectroscopy

Advantages of the Method

Raman spectroscopy has a number of indisputable advantages which appreciate scientist form variety spheres. These benefits go towards the growth of popularity of Raman spectroscopy in laboratories worldwide. The method is:

- Non-destructive. Sample can be after Raman analysis consequently treated by other procedures. Non-destructiveness also enables potential as in-vivo diagnostic tool in medicine providing information about both the chemical and morphologic structure of tissue in quasi-real time.
- Non-contact. No contamination of a sample happens. This is convenience and safe. For example dangerous and toxic samples or those with unpleasant smell, substances with unknown composition, properties, and history from the scene of an accident, samples, that are unstable to air or moisture can be measured through protective or covering layers or packages from other materials as glass or polymers. Spectra of these packages can be later subtracted using the software designed for manipulation with data. Raman spectroscopy therefore appears as an easy and strikingly useful technique analysing and identification of materials even in packing.
- No need for sample preparation prior the analysis is required - that is convenient and prompt. No physical or chemical conditioning of sample (diluting, pressing into pellets, etc.) is necessary.
- Highly sensitive. High spatial resolution is in the order of micrometers. Currently the most commonly used Raman spectrometers are combined with microscopes. Then only very small volume (about ones of μm in diameter) of a sample is needed for collecting Raman spectra when using Raman microscopy. This interconnection yields many benefits e. g. for forensic science investigation of trace amounts of evidences.
- A rapid method. High quality Raman data are acquired in very short periods of time, often within seconds. E. g. chemical analyses generally take minutes or even hours. This kind of analysis definitely spares time. The rapidity of the method yields further potentialities

inherent in a scope of almost immediate response and quasi real time processes e. g. chemical changes kinetics monitoring.

- Applicable to a wide range of substances. Via Raman spectroscopy is possible to measure liquids, transparent solids and gases. Samples can be in a form of powder, crystal set, fiber, thin layer, gel, solution, etc. with no concern with sample size, shape or thickness. The method is also appropriate for analyses of organic and inorganic compounds.
- Providing possibility of aqueous solutions exploration since water is very weak Raman scatterer and generally almost does not interfere with Raman spectral analysis. This is a great advantage in comparison with infrared spectroscopy where the presence of water or moisture is unfavourable.
- Granting highly specific chemical "fingerprint". Each compound gives rise to a unique Raman spectrum.
- The standard spectral range ideal for both organic and inorganic samples covers from 100 cm^{-1} to 3200 cm^{-1} and can be covered by a single spectra recording.
- The intensity of spectral features is directly proportional to the particular species concentration.
- Remote analyses can be realized. Transmission of laser light and Raman scattered signal can be done by fibre optic probes over long distances – up to hundreds of meters far from the base of Raman analyser.

Drawbacks of Raman Spectroscopy

As with all problems we solve and take into account related advantages, we have to consider either the weak side - limitations. In spite of many advantages of Raman spectroscopy a well-known competing process can appear along with the Raman scattering: fluorescence. Fluorescence signal is generated by a part of matter that undergoes an electronic transition to an excited state what initiate emission of light at a wavelength which is changed from the excitation laser. A mechanism controlling both effects is similar and determines that if one of the phenomena occurs, the second will likely as well. Fluorescence initiation probability versus probability of Raman scattering initiation is in the order of 1 photon in 10^3 – 10^5 versus 1 photon in 10^6 – 10^9. Hence we always have to consider fluorescence interference. Raman spectra, for instance, of certain biological samples are often overlaid by fluorescence when visible wavelengths of laser are used. There exists ways to avoid or at least minimize this adverse effect A solution often consists in a selection of suitable laser wavelengths preferably with lower photon energy. As the laser excitation wavelength gets shorter the Raman scattering intensity increases but fluorescence tends to be less troublesome as the laser wavelength gets longer. Thereby spectrometers equipped by several lasers with different excitation wavelengths are recommended to overcome the problem with fluorescence. The interfering luminescence background can be also in some cases reduced by so called "bleaching", i.e. prolonged sample illumination with the laser beam antecedent to concrete measurement.

Even though Raman spectroscopy can be applied on measuring an extend spectrum of materials, metals and alloys are not Raman active. Other disadvantage of Raman spectroscopy is the sensitivity of the method. In spite of the fact, that the particularity of Raman spectroscopy is remarkable, the conversion efficiency of Raman effect is rather poor since only a scarcity (about 10^{-6}) of initial photons is non-elastically scattered. In other words highly sensitive equipment (lasers, detectors, filters), what usually means also quite expensive, is demanded.

One more disability is eventual damage or laser-induced degradation of a sensitive or light absorbing sample (especially when examining unknown matter) through intensive laser radiation however experienced analysts can anticipate this complication in the right preference of measurement procedure.

Applications

Raman spectroscopy has been reborn in recent years. As the more sophisticated spectroscopic and computer technology access increased, Raman spectroscopy has become a tool for not only purely research, but also for contrivable routine industrial analyses. These days, Raman spectroscopy finds applications, which number is still rising, in variety scientific and industrial disciplines and branches such as:

- Material sciences including semiconductor industry, nanotechnology, solid state physics, synthetic polymer materials and others. Raman spectroscopy is one of the best-known methods, it is of central importance for all the specification and structural analysis of almost all kinds of materials (amorphous, partially crystallic, transparent, non-transparent samples, samples with different surface textures).
- Nanotechnology: Raman spectroscopy is suitable technique for characterization of nanomaterials as nanocomposites, nano-sized crystals, polymers and semiconductors developed in a form of films, wires or dots or modern and ancient ceramics. The method provides determination of nanocrystals, chirality, semidiameters in nanomaterials, also characterizing micromechanical behavior or synthesis processes control (polymer curing, laser ablation, electrochemical deposition). It is a good tool for mapping and probing nanophases dispersed in matrix.
- Semiconductor industry: Raman spectroscopy allows semiconductor impurities determination in silicone substrates and diamond-like carbon coatings (a point measurement on silicone can be obtained in about 0.1 second), identification of defects particles on the material surfaces. Such results notably affect device yields and the economics of the process line.
- Solid-state physics: Raman spectroscopy systems find application in material characterization, finding the crystallographic orientation, etc.
- Synthetic polymer materials: Raman spectroscopy serves for real-time monitoring of polymerization reactions for the purpose of controlling the processing time, for plastic identification for recycling purposes, measuring the thickness of polymers such as protective and coating films, evaluation of polymer material under applied strain or quality control of incoming/outgoing products.
- Chemistry: Raman spectroscopy provides a chemical fingerprint for identification of a

molecule, since vibrational information is specific to the chemical bonds and symmetry of molecules. Except for identification the method is used for characterization and analyses of organic and inorganic substances, including carbon materials, solvents, films and for chemical processes monitoring.

- Forensic sciences: Forensic scientists often deals with really unknown samples that cannot be reproduced, require multiple analyses, but their amount is limited. Raman spectroscopy/microscopy is becoming a tool of major importance in forensic science since it satisfies most of forensic examination criteria, as are mentioned above in advantages: it is a non-destructive, non-contact method without the necessity of sample preparation and possibility of application on a wide range of materials. In-situ measurements can be realized, meaning no contamination of evidences during taking samples. Raman microscopy brings benefits of optical microscopy, what means that the sample can be surveyed under microscope, particles or locations isolated and consequently their Raman spectra acquired.

Method is used for forensic analyses and rapid identification of trace amounts of substances in evidential materials as are paints, inks from documents, pigments, explosive particles, inflammables, drugs, illegal active ingredients, fibers, gunpowder residues, chemical and biological agents, plastics and other various forensic evidences. An example of Raman spectra of cocaine is shown in Figure. In most cases forensic examination comprise comparative analyses, hence the spectra libraries and databases are required. In case of having comparable samples it is easy to demonstrate the match or identify counterfeit for instance of documents or historical artworks.

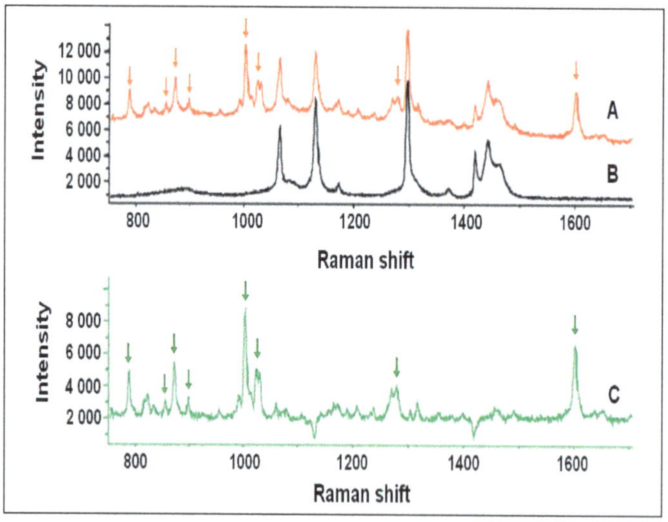

A – Raman spectrum of cocaine in plastic container, B – Raman spectrum of plastic bag, C – the difference spectrum of cocaine obtained by subtraction of spectra A-B.

- Security forces, fire brigades: Identification of unknown or hazardous substances, by instance detection of explosives or drugs. These units often work with portable spectrometers enabling identification right in action. Appreciable benefit of method is potentiality of measuring samples through containers.

- Pharmaceutical industry: The role that Raman spectroscopy plays in pharmaceutical research, development and manufacturing is still rising. Raman analysis is an efficient tool for control of quality and purity of pharmaceuticals, active substances and excipients (even through packing). Counterfeits pharmaceutical signify a health risk since they may contain harmful impurities, wrong ingredients or incorrect amounts of active pharmaceuticals. Raman spectroscopy has been also successfully used for adulterated pharmaceuticals revealing.

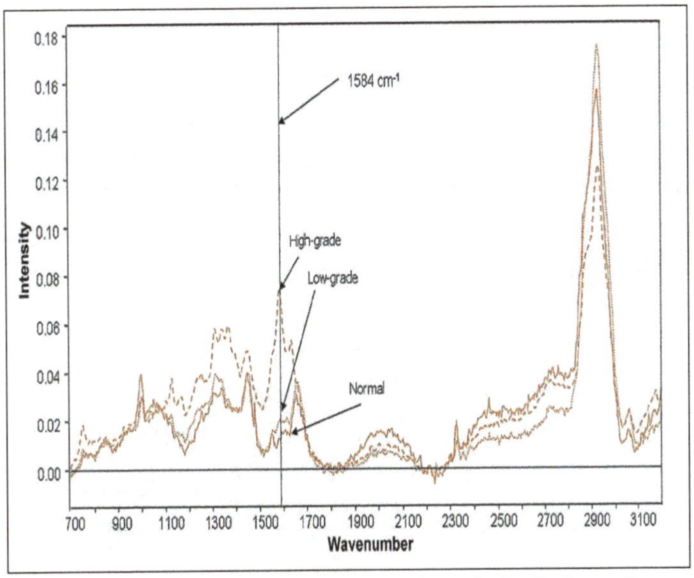

Raman spectrum of high-grade urothelial carcinoma (UC; dashed) tissue, low-grade UC tissue (dotted), and normal tissue (solid) – the notable peak at 1584 cm^{-1}.

- Medicine and biology: Applications benefiting Raman spectroscopy involve e. g. DNA analyses, prognoses and diagnoses of carcinomas (example displayed in Figure), measuring blood and tissue oxygenation, study of biological systems. With regard to non-destructiveness and contactless the great potential of the method for invivo medical examination is studied.

- Food Product: Raman spectroscopy serves for detecting bacteria and contaminants in food product or measurements of fatty acid unsaturation in food oils and other ingredients.

- Geology and mineralogy: Raman spectroscopy serves for identification of the principal mineral phases or classification of rocks, etc.

- Art: The main intention of any analytical studies of any ancient artwork should be to get as much information as possible by non-destructive methods. Raman/MicroRaman spectroscopy is suitable and effective for this purpose. Raman examination of artworks and artefacts (the most often are paintings and pottery) reveals worthy information for conservators or those of general historical interest. Knowledge of artistic materials in connection with the time period or the location in particular region and also about the artists' modus operandi can be gained.

Followed issues are solved at the present time using Renishaw in Via Basis Raman microscope (schema in Figure) with the 514 nm excitation Argon ion laser at Faculty of Applied Informatics, Tomas Bata University in Zlin, Czech Republic.

- The methodology for the detection of small hexavalent chromium concentrations with the use of Raman spectroscopy is developed. Chromium is an element of a common occurrence in nature. In living environment can be found in several modifications. The most frequent form is metallic chromium Cr(0), trivalent chromium Cr(III) and hexavalent chromium Cr(VI). Trivalent and hexavalent chromium compounds are produced in large quantities and are accessible to most of the population. However, Cr(VI) is a carcinogenic substance and may cause health risks. There is a possibility of conversion of Cr(III) into Cr(VI). It is already proved that Raman spectroscopy can distinguish Cr(III) from Cr(VI) [10]. Obtained results showed the complication with fluorescence that masks the Raman spectra of chromium compounds contained in natural polymers. This situation has been so far solved by bleaching prolonged illumination of the sample before own measurement, but without success. On the other hand, shortening the exposure time to tenths of seconds allowed recording of the Raman spectra.

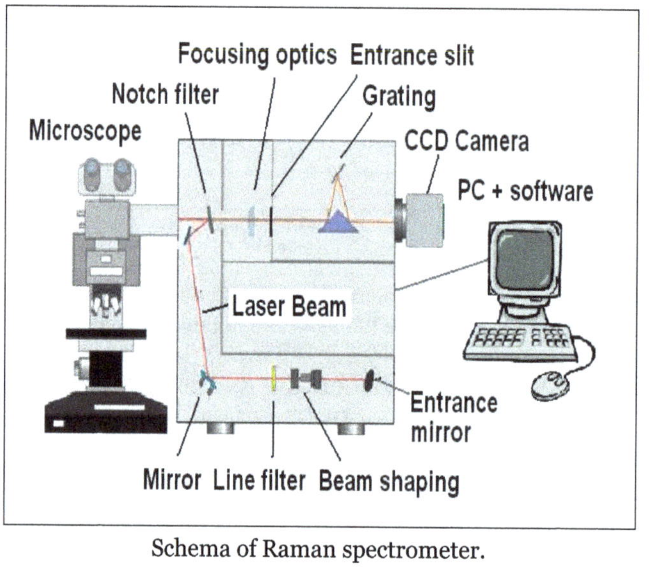

Schema of Raman spectrometer.

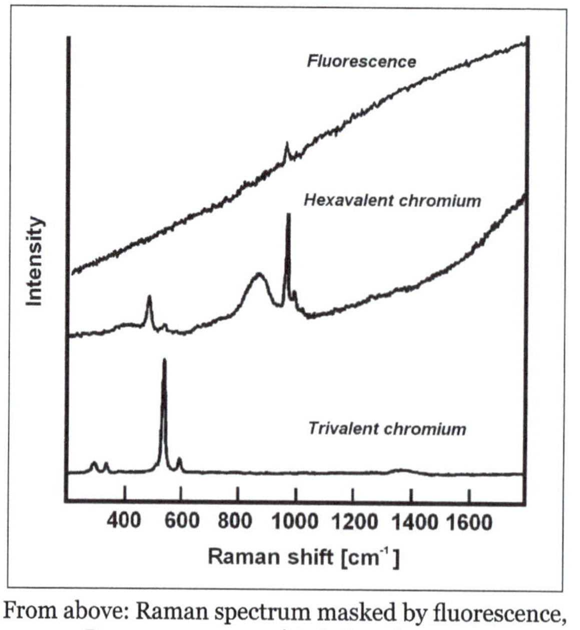

From above: Raman spectrum masked by fluorescence, Raman spectrum of Cr(VI) and Cr(III).

Fibers of leather with linked hexavalent chromium.

- Raman spectroscopy has been applied on monitoring of the curing process of epoxy resins. Epoxy resins are one of the most versatile polymers with a number of good to excellent properties and owing to them are epoxy based materials intensively applied in many technical areas and industries. Due to a time series measurement can the kinetics of crosslinking be controlled and the changes in active chemical groups and bonds recorded and analysed. The progression of crosslinking process dependence on temperature has been also studied. In Figure is displayed 3D map of crosslinking reaction at the temperature 25 °C recorded over 30 minutes with evident peak corresponding to gelation time.
- Application of Multi-Walled Carbon Nanotube (MWCNT) for the purpose of conducting layers creation is investigated. Raman spectroscopy is very useful in the detection of carbon form in fundamental research and also in industrial usage. This technique is able to distinguish affect properties of electric conductivity of different forms of carbon, because of evident Raman spectral variances, and also map them. While, for example, SEM indicates only presence of elemental carbon in the sample. Different concentrations of MWCNT are used for shielding electromagnetic fields or elimination of static charge and others.
- Determination of inks within the scope of security usage is an upcoming problematic at present time dealing with the possibility of revealing ink document falsifications. Ten different blue ball-pen inks with the known producer were examined for the purpose of ink identification and aging and eventual document dating. Actual results show very similar Raman spectra at the inks from the same producer e.g. KOH-I-NOOR, other are comparable with a few differences, which, however, should be sufficient for the identification. Two samples were sheltered by strong fluorescence and despite different setting of measuring parameters, Raman spectra was impossible to obtain. This can be after all also a sort of hint for identification following by other methods.

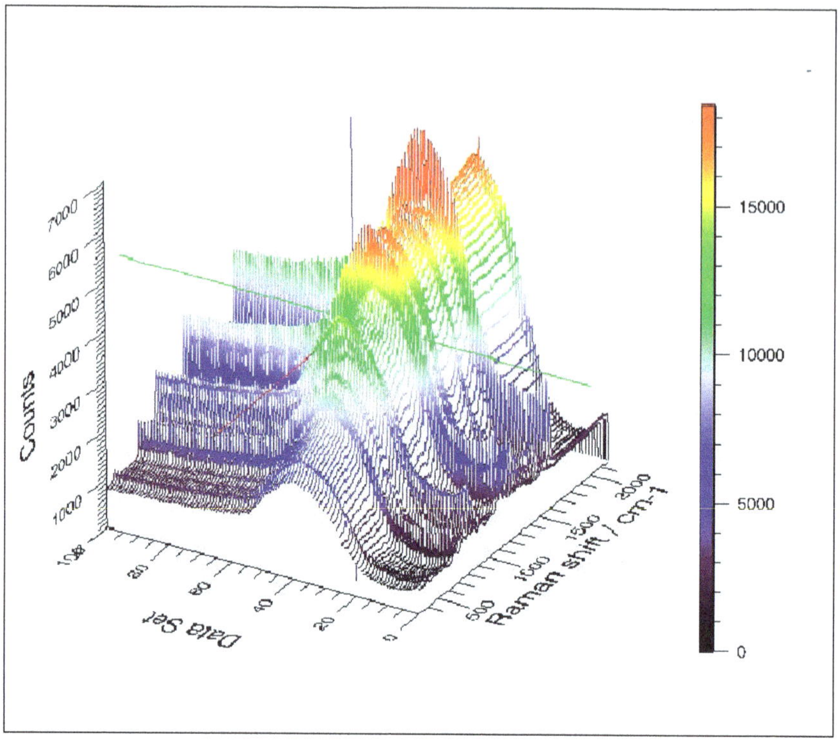

3D map of epoxy resin crosslinking process.

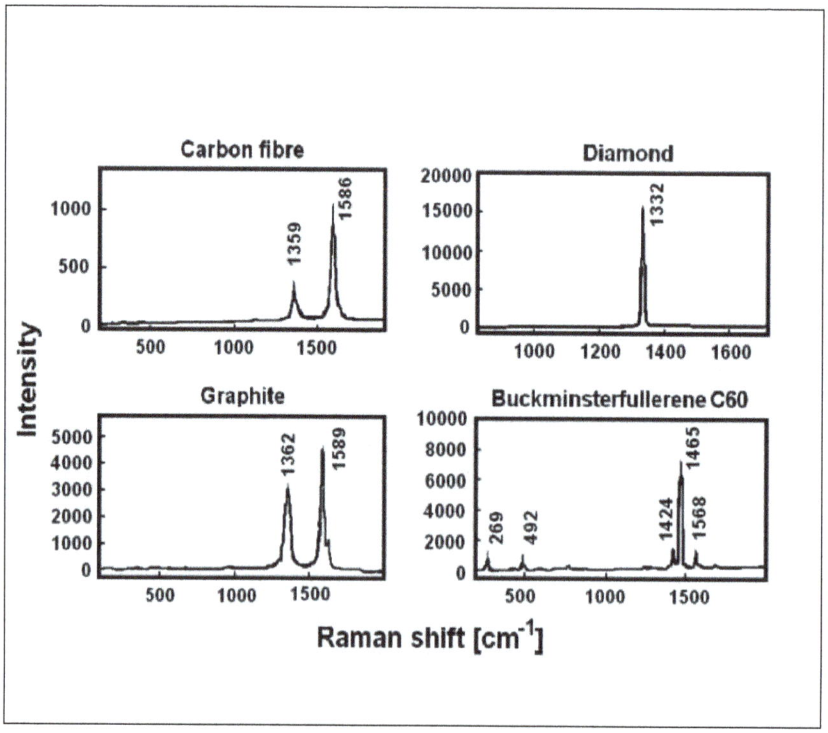

Raman spectra of different modifications of carbon.

- Another application deals with the mapping of the distribution of active substances in pharmaceuticals. In Figures are for illustration shown 3D maps of distribution of acetylsalicylic acid and ascorbic acid in a Aspirin.

Other applications are prepared in present time.

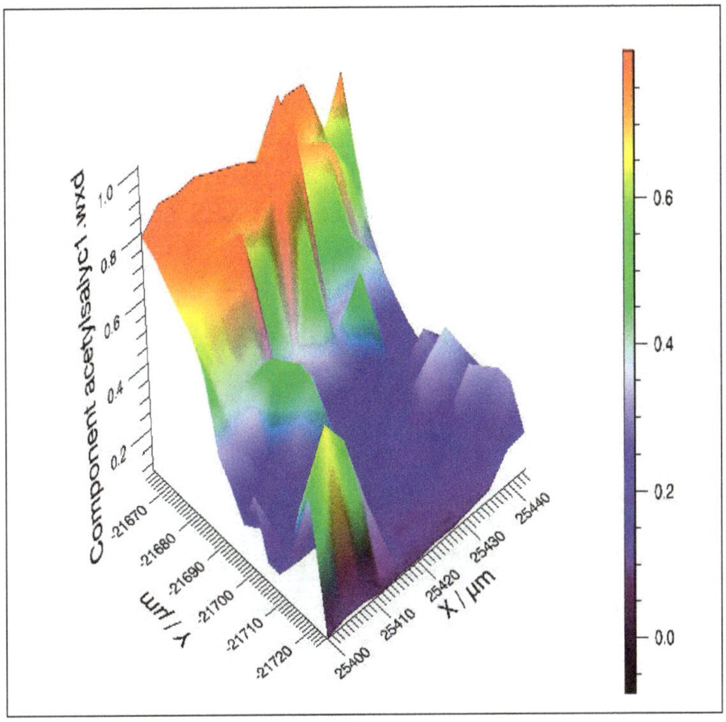

3D map of acetylsalicylic acid in Aspirin.

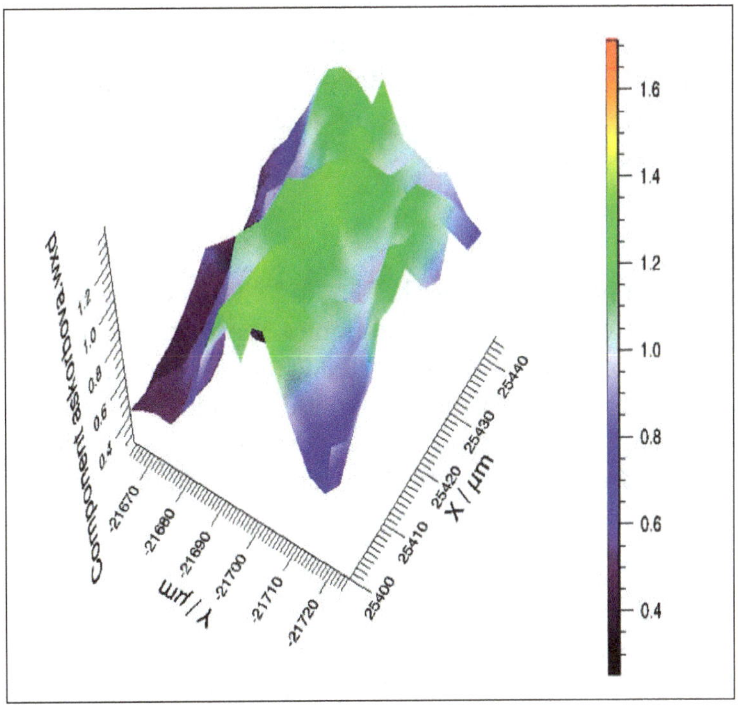

3D map of ascorbic acid in Aspirin.

X-Ray Photoelectron Spectroscopy

X-ray photoelectron spectroscopy (or XPS) belongs to the group of surface analysis methods wherein the sample under investigation is bombarded with photons/electrons/ion/etc. for exciting the emission of photons/electrons/ion/etc.

XPS involves irradiating the sample with low energy ((~1.5 keV) X-rays such that photoelectric effect is induced. An electron spectrometer with high resolution records the energy spectrum of emitted photoelectrons. Highest available vacuum conditions are required for XPS experiments. High vacuum assists photoelectron transport to the analyser, and also avoids recontamination of clean samples. Contamination is an important concern in XPS, since it is a surface sensitive technique and the sampling depth is usually a few nm.

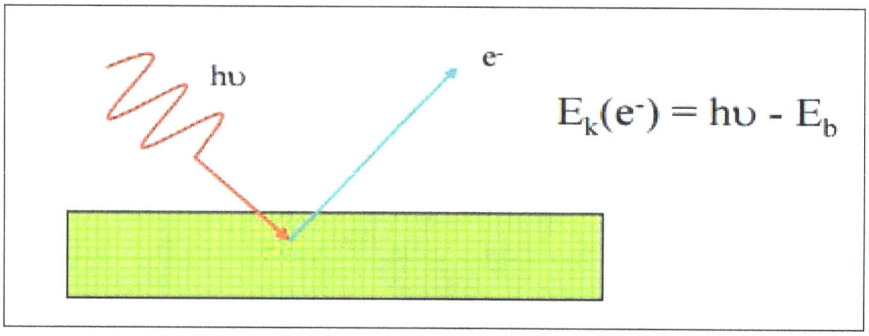

Principle

Surface analysis by XPS involved irradiation of the sample by low-energy (and mono-energetic) x-rays and the subsequent analysis of the energy of emitted electrons. Typically used x-rays are Kα lines of Mg (1.2536 keV) and Al (1.4866 keV). The penetration depth of these photons in solids is limited to a few microns. Thus, interactions take place between the incident photons are the surface atoms leading to the photoelectric emission of electrons. The kinetic energy (K.E.) of the emitted electrons is expressed as:

$$K.E. = h\nu - B.E. - \varphi_s$$

Here hν is the energy of the photon, B.E. is the binding energy of the atomic orbital from which the electron is released, and φ_s, is the work function of spectrometer. B.E. can also be described as the energy difference of the initial and final states when an electron is released from the atom. Electrons emitted from an ion possess different kinetic energies owing to the possibilities of different final states of the ion from each type of atom. Besides, there exists different probability or cross-section for each of the final state. Figure shows relative B.E. and ionization cross sections for an atom.

Binding Energy

In a photoemission process, the total energy of the system, before emission, is equal to the sum of the energies of incident X-ray photon (hυ) and target atom in its initial state (E_i). After emission, total energy of the system equals the sum of the K.E. of the emitted electron (E_k) and energy of ionized atom in its final state (E_f). Equating the total energy before and after emission and comparing with Einstein's equation gives the B.E. as the energy difference between the final and initial states of the target atom (i.e., $E_f - E_i$).

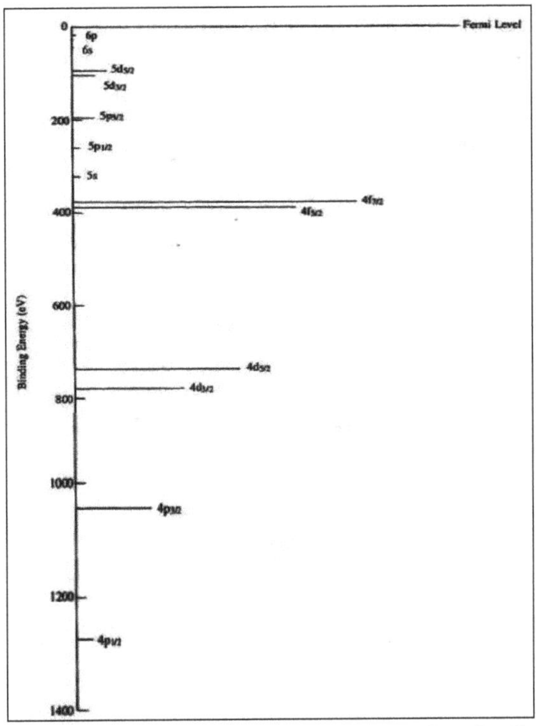

Relative B.E. and ionization cross-section for U. The B.E. is proportional to the distance below the line indicating the fermi levels, and the ionization cross section is proportional to the length of line.

In figure, Fermi level corresponds to zero B.E. and the depth below it reflects the relative energy of the ionized atom after emission, or simply B.E. of electron. Relative probabilities of different ionization processes are indicated by the length of the lines. Ionization causes splitting of the levels p, d and f, creating vacancies in $p_{1/2}$, $p_{3/2}$, $d_{3/2}$ $d_{5/2}$, $f_{5/2}$ and $f_{7/2}$. The splitting ratio of spin-orbit is 1:2 for p, 2:3 ford and 3:4 for f levels. For instance, Figure shows the spin-orbit splitting in Si 2p level.

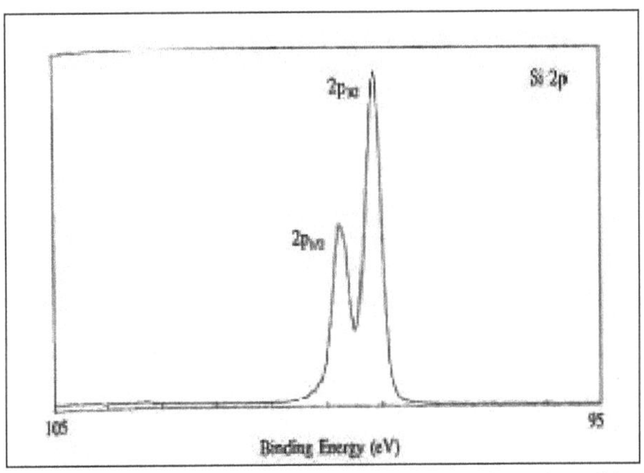

High resolution spectrum of single crystal Si indicating the spin orbit splitting of 2p.

XPS analysis relies on the fact that each element possesses a unique set of B.E. Thus, the concentration of elements at the surface can by identified and determined. B.E. of elements vary (the chemical shifts) because of differences in the chemical potential and polarizability of compounds. These shifts lead to the identification of chemical state of the material under investigation.

The excited ions left after photoemission, may also eject Auger electrons. Auger emission typically occurs after 10^{-14} s of the photoelectric emission. Another competing emission of a fluorescent X-ray photon may also take place. Schematic illustration of Auger phenomenon is shown in Figure, where an electron from outer shell occupies the vacancy in the inner orbital by emitting a second electron in the process. This second electron carries the extra energy, and is called the Auger electron. The K.E. of Auger electron is the energy difference between the initial ion and the doubly charged final ion. Additionally, the K.E. of Auger electron does not depend on the mode of initial ionization. Therefore, photoionization often generates two electrons: one photoelectron and another Auger electron. The sum of the K.E. of the emitted electrons cannot be larger than that of the ionizing photons.

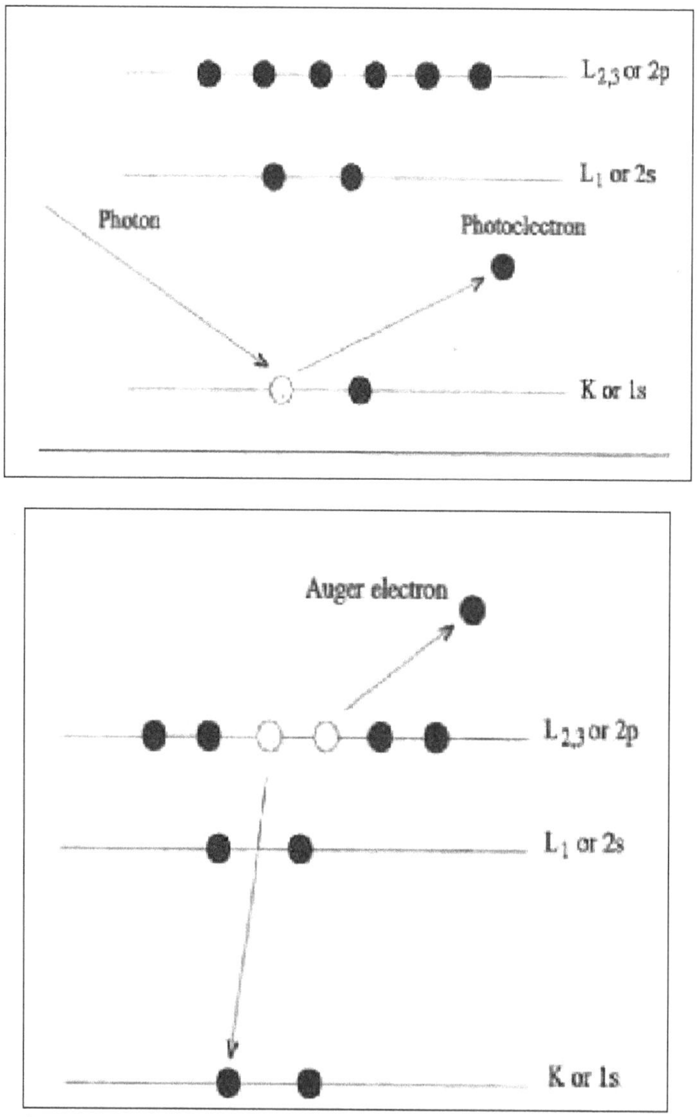

Top: XPS emission for a hypothetical atom where an incident photon ejects a photoelectron. Bottom: relaxation of the atomemitting a $KL_{23}L_{23}$ electron. The simultaneous two electron coulombic rearrangement results in a final state with two electron vacancies.

Electrons undergo far more interactions with matter than photons, this result in extreme short path length of electrons (~ few angstroms) in comparison to photons (~ few microns). Therefore when the depth of ionization is of the order of few microns, the electrons originating from deep into the bulk of solid (i.e., at a distance more than few angstroms from the surface), experience high energy losses. However, the electrons originating from the surface (or up-to a depth of few angstroms from the surface) do not lose their energy, and cause peaks in the XPS spectrum. These electrons are crucial in XPS analyses. The deeper electrons experience elastic loss process before releasing from the background. Figure indicates the calculated inelastic mean free path (IMFP) of electrons in different materials.

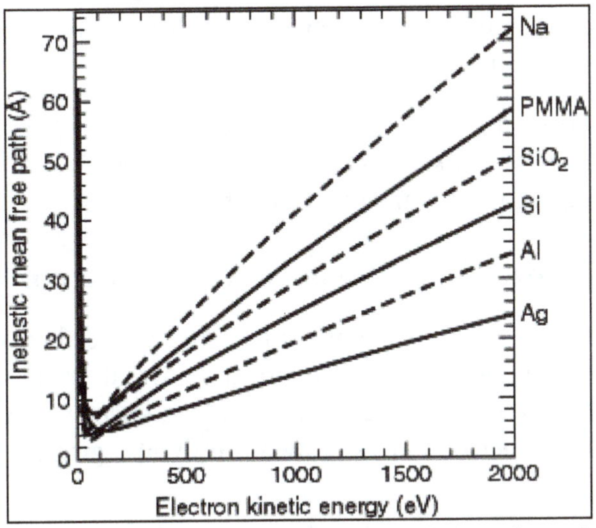

Calculated IMFP for electrons in different metals from the methods of S. Tanuma, C.J. Powell and D.R. Penn, Surf.

An electron spectrometer detects the electrons emitted from the sample as a function of their K.E. The analyser works as an energy window, termed as pass energy, and accepts only the electrons with energy within the range of this window. The pass energy is usually fixed for maintaining a constant energy resolution. Incoming electrons are adjusted to the pass energy before entering the energy analyser. A variable electrostatic field is used before the analyser in order to scan the different energies. This is called retardation voltage and can be varied from zero to more than the energy of the photon. Electrons are detected as discrete events, and the number of electrons for a given detection time and energy is stored and displayed.

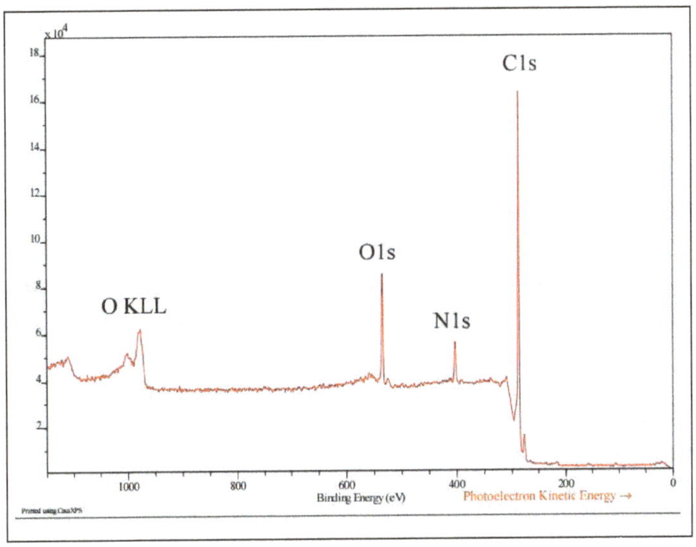

XPS spectrum of polystyrene.

Figure shows XPS spectrum of polystyrene sample which was pre-treated with nitrogen plasma. The spectrum was recorded by exciting the sample by Al Kα rays. The B.E. scale is drawn from right to left and thus, the K.E. of the photoelectrons increase in reverse direction.

The spectrum contains three photoemissions (electrons emitted from 1s orbitals of C, N, and O atoms in the sample surface), and one Auger emission (O KLL). The background of these peaks results due to the electrons excited by X-ray Bremsstrahlung radiation at low B.E., and from in elastically scattered photoelectrons at higher B.E. (to the either side of C-1s peak).

The energy of Auger electrons does not depend on the energy of X-ray. Thus, whenever Auger peaks overlap with photoelectron peaks, Auger peaks are displaced on the B.E. scale by altering the X-ray energy (such as using Mg Kα instead of Al Kα, etc.).

Factors Affecting the Intensity of Peaks in the Spectrum

Consider a volume element of the sample with thickness dz at a depth z under its surface. The photoelectrons emitted at an angle θ with respect to the normal to the sample surface (the "photoemission angle") enter the detector and contribute to the spectrum. Usually, X-rays' reflection and refraction induced photoelectrons are significant only at a grazing angle $\varphi < 5^{0}$, and may be ignored in other cases.

1) The intensity of the x-rays at depth z

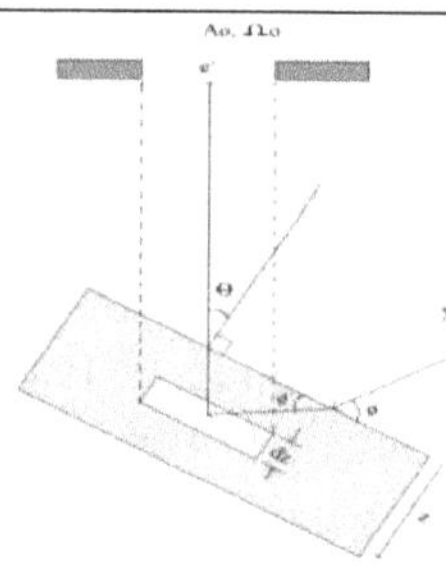

$$\gamma(1-r)\frac{\sin\phi}{\sin\phi'}e^{\frac{-z}{\lambda_{h\upsilon}\sin\phi'}}$$

γ = incident X-ray flux
r = coefficient of reflection
λ_{hv} = attenuation length of X-ray photons

X-ray flux is not usually known. Therefore, this factor is not considered and peak intensity ratios or composition of the sample in terms of atom percent (considering one peak from each chemical element) is taken into account. The extent to which volume element generates counted photoelectrons is influenced by the particular geometry of the experimental setup. The following cases may normally arise:

- More surface is irradiated by X-rays than detectable by the analyser optics (e.g., a non-monochromated X-ray source may have been used).
- Less-surface is irradiated by X-rays than observed by the analyser optics (using a monochromated X-ray source).

Both these cases must be considered carefully and sometimes a transition between the two cases may be observed when there is a change in photoemission angle θ. However, these geometrical factors can be cancelled by evaluating sample composition as atomic percentage.

The analyser acceptance angle Ω° is a function of the lens programs and aperture settings, and is unknown. This is also cancelled by evaluating sample composition as atomic percentage(considering peak intensity ratios).

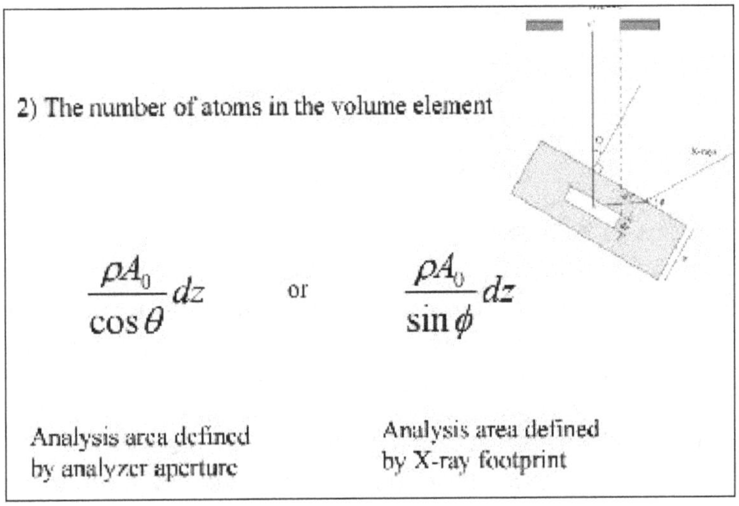

The attenuation of emitted photoelectrons flux gives XPS its surface sensitivity. The inelastic mean free path is the average distance between collisions where the photoelectron loses its energy. If a photoelectron loses K.E. in a collision, it will contribute to the background rather than to the characteristic "no-loss" peak in the spectrum.

Instrumentation

The instrumentation of XPS system can be explained using the experimental set up shown in figure. In a typical set up, when the sample kept in ultra-high vacuum is illuminated by the photons with energy (hv); soft x-rays, then after direct transfer of energy from the photons to the core level electrons their atoms on the surface emit electrons or so called photoelectrons. The process can be explained as follows:

- A bound electron adsorbs a photon and converts part of the energy to kinetic energy.
- While leaving the atom, some energy of the electron is consumed in overcoming the Coulombic attraction of nucleus, thereby decreasing its kinetic energy.
- At this time the outer orbitals readjust, reducing the energy of the final state which is being created and delivering this extra energy to the outgoing electron.
- Then the XPS spectrum is obtained by determining the K.E. and number of electrons escaping from upper 0 to 10 nm of sample under investigation.
- XPS may be routinely employed to study inorganic compounds including metal alloys, semiconductors, polymers, glasses, ceramics, ion-modified materials, etc.

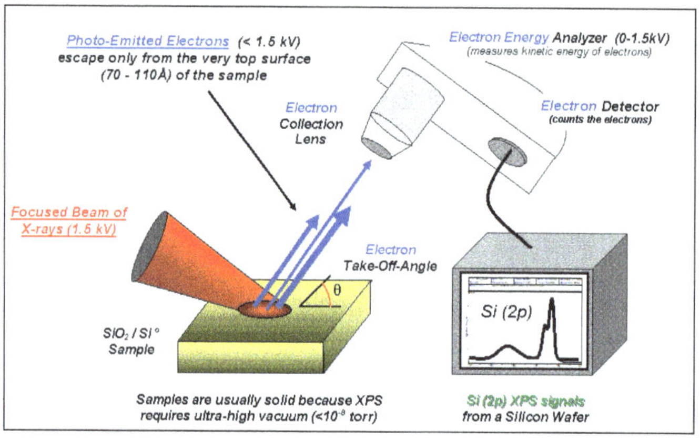

Monochromatic XPS system.

There are many limitations to XPS that are not based on the samples or preparation, but on the machine itself. One such limitation is that XPS cannot detect hydrogen or helium. This, of course, leads to a ratio of elements in the sample that is not entirely accurate, as there is always some amount of hydrogen. It is a common fallacy to assume the percent of atoms obtained from XPS data are completely accurate due to this presence of undetected hydrogen.

It is possible to indirectly measure the amount of hydrogen in a sample using XPS, but it is not very accurate and has to be done in a roundabout, often time consuming manner. If the sample contains hydrogen with a partial positive charge (i.e. OH), the sample can be washed in sodium naphthalenide ($C_{10}H_8Na$). This replaces this hydrogen with sodium, which can then be measured. The sodium to oxygen ratio that is obtained infers the hydrogen to oxygen ratio, assuming that all the hydrogen atoms have reacted.

XPS can only give an average measurement, as the electrons lower down in the sample will lose more energy as they pass other atoms while the electrons on the surface retain their original kinetic energy. The electrons from lower layers can also undergo inelastic or elastic scattering, seen in Figure. This scattering may have a significant impact on data at higher angles of emission. The beam itself is also relatively wide, with the smallest width ranging from 10 – 200 μm, lending to the observed average composition inside the beam area. Due to this, XPS cannot differentiate sections of elements if the sections are smaller than the size of the beam.

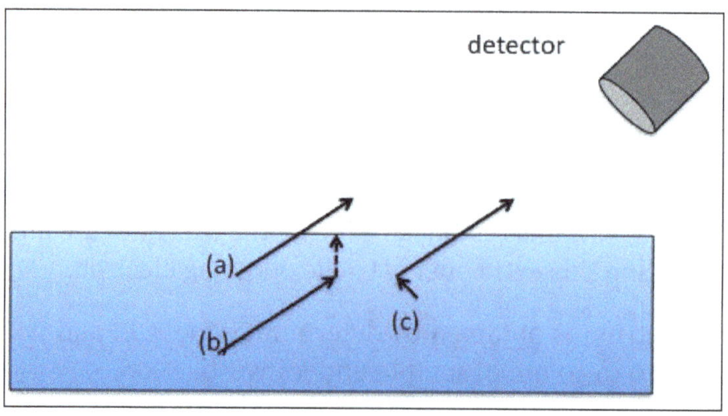

Schematic representation of (a) no scattering, (b) inelastic scattering, and (c) elastic scattering.

Sample reaction or degredation are important considerations. Caution should be exercised when analysing polymers, as they are often chemically active and X-rays will provide energy to start degrading the polymer, altering the properties of the sample. One method found to help overcome this particular limitation is to use angle-resolved X-ray photoelectron spectroscopy (ARXPS). XPS can often reduce certain metal salts, such as Cu^{2+}. This reduction will give peaks that indicate a certain set of properties or chemical environments when it could be completely different. It needs to be understood that charges can build up on the surface of the sample due to a number of reasons, specifically due to the loss of electrons during the XPS experiment. The charge on the surface will interact with the electrons escaping from the sample, affecting the data obtained. If the charge collecting is positive, the electrons that have been knocked off will be attracted to the charge, slowing the electrons. The detector will pick up a lower kinetic energy of the electrons, and thus calculate a different binding energy than the one expected, giving peaks which could be labeled with an incorrect oxidation state or chemical environment. To overcome this, the spectra must be charge referenced by one of the following methods: using the naturally occurring graphite peak as a reference, sputtering with gold and using the gold peak as a reference or flooding the sample with the ion gun and waiting until the desired peak stops shifting.

Limitations with Surfactants and Sputtering

While it is known that sputtering is destructive, there are a few other limitations that are not often considered. The beam of X-rays is relatively large, giving an average composition in the analysis. Sputtering has the same limitation. If the surfactant or layers are not homogeneous, then when the sputtering is finished and detection begins, the analysis will show a homogeneous section, due to the size of both the beam and sputtered area, while it is actually separate sections of elements.

The chemistry of the compounds can be changed with sputtering, as it removes atoms that were bonded, changing the oxidation state of a metal or the hybridization of a non-metal. It can also introduce charges if the sample is non-conducting or supported on a non-conducting surface.

Auger Electron Spectroscopy

Auger electron spectroscopy (AES) is a non-destructive electron spectroscopy technique for partial identification of elemental composition of materials. The samples can include surface, thin film, interface, etc. High surface sensitivity (analysis depth < 100 Å) and a relatively lower detection limit (~0.1 atomic percent) make AES a very popular technique for surface analyses. AES can also be used to differentiate between elements which may be adjacent in periodic table. Additionally, it can be used for almost all elements, i.e., from lithium to uranium and beyond. AES also demonstrates atomic number dependent sensitivity which may vary by up-to one order of magnitude. Bonding or chemical state information can be obtained from the chemical shift and line shape observed in AES. However, AES is less accurate than XPS for obtaining bond information. AES has depth resolution of 5–25 Å, which makes it particularly attractive for depth profiling with simultaneous ion sputtering. The lateral resolution of AES (less than 100 Å) is much better than XPS. Thus, nanostructures can be readily imaged along with creating 2-D maps of elemental composition of surface via scanning Auger microscopy (SAM). AES is a rather fast technique, despite being an expensive and sophisticated technique.

Auger is a three-electron process, wherein an electron beam (having 1 to 10 keV energy) is bombarded onto a solid surface. The impact of the beam ejects an electron from inner or core shell of the surface atom, thereby creating an ionized excited atom. An electron from an outer shell can occupy the vacancy produced in the core-shell, via a non-radiative transition. The excess energy of the electron is removed via:

- X-ray emission, this phenomenon forms the basis for X-ray fluorescence (XRF)/electron microprobe (EMP) analysis.
- Emitting another electron from the atom. These atoms are termed as Auger electrons, after Pierre Auger, discoverer of this process in 1920s. Around the same time Lise Meitner also invented this effect in an independent investigation.

The X-ray emission of X-rays (i) creates singly ionized atoms, whereas the atoms in Auger process (ii) become doubly ionized. Interestingly, AES cannot be used to detect hydrogen and helium, as Auger involves three electrons. Further, even though lithium possesses three electrons, an individual lithium atom in its ground state does not produce Auger peaks because the atom has only two energy levels which have electrons. Nonetheless, multiply excited lithium atoms have demonstrated Auger peaks. Additionally, electrons present in valence band of solid lithium facilitate Auger transitions of type KVV.

The low energy Auger electrons have a short mean free path which is responsible for the high surface sensitivity of this technique. Even though the atomic excitations can occur upto the depths of around 10,000 angstroms beneath the surface, it is only from the depth of ~100 angstroms that Auger electrons can be emitted without losing energy. On the contrary, electron microprobe analysis (EMA), where in X-ray photons are detected, is rather a bulk technique. The electronic states related to Auger process have very short lifetimes, thus, according to Heisenberg's uncertainty principle, the peaks in the spectrum are usually wide (~1 to 2 eV). Auger electrons' kinetic energy (K.E.) is specific to the emitting atom. Consequently, particular elements can be identified by measuring this K.E. of the emitted electrons in Auger process. Since the K.E. of electrons in Auger process is determined by the binding energy (B.E.) of electronic levels, any shift in K.E. yields detail about the oxidation state and bonding environment of the atoms present at the surface. Additionally, quantitative details can also be obtained from the Auger peak intensities.

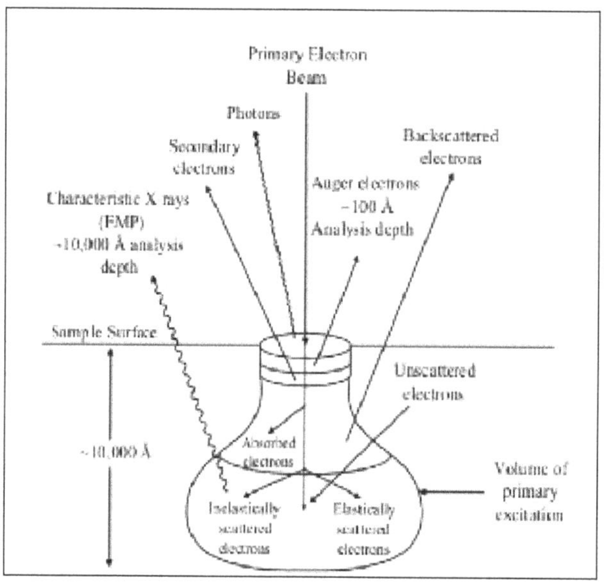

Depiction of different locations and result of electron-solid interactions.

The number of electrons detected is plotted against the K.E. of electrons to obtain direct Auger spectrum. In addition to this, for distinguishing small peaks in the spectrum, usually the first derivative of the number of electrons is plotted against their K.E. The derivative Auger spectrum serves two purposes: Auger peaks are enhanced; and the background produced from secondary and backscattered electrons is suppressed. Ultra high vacuum chamber is required for AES investigations, such that electron beam can travel freely and contamination from atmosphere is avoided.

Principle

Figure depicts the schematics of Auger phenomenon is solids. The figure shows KL_2L_3 Auger transition which involves ionization, relaxation, and emission, and is summarized below:

- Ionization: Highly energetic incident electrons remove an electron from the core shell of the atom, leading to the formation of a vacancy in K-shell. Thus, an electronically excited ion is produced.
- Relaxation: The K-shell vacancy created during previous step is quickly filled by an electron coming from L_2 shell via non-radiative transition.
- Emission: An electron (Auger electron) is emitted from L_3 shell, thereby releasing the extra energy of the excited ion.

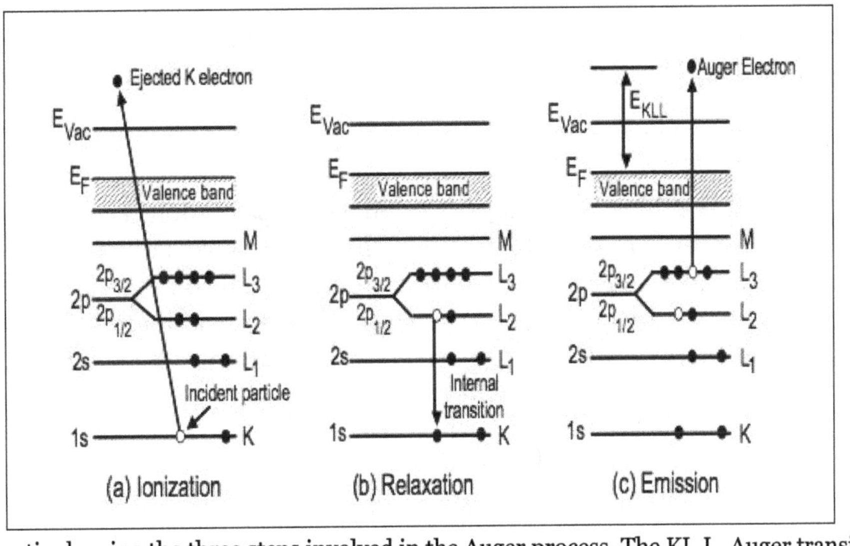

Schematic showing the three steps involved in the Auger process. The KL_2L_3 Auger transition is illustrated. The open circles symbolize holes (absence of electrons).

Auger transitions are named on the basis of the energy levels involved in the process, e.g., the transitions involved in the above figure is named KL_2L_3. In certain cases, such as atoms having high atomic numbers, many Auger transitions are available. However, most of them have low probability of occurring. Further, some of the energetically allowed transitions are also forbidden by the selection rules. If a valence electron takes part in the Auger transition, letter V is used (such as KLV, KVV, etc.). Transitions involving core level electrons use letter C (such as CVV). Transitions of ABB (such as KLL, LMM) type are strongest. Another strong transitions are AAB type (also called as Coster-Kronig) transitions. Emission of X-rays is a competing phenomenon for Auger process since the energy difference between the core and outer levels can equally be released in the form of a characteristic X-ray. Therefore, the sum of the yields of Auger emission and X-ray emission is one. AES typically operates in the energy range where the probabilities of Auger emission are much higher than that for X-ray emission. In addition to this, the cross-section for Auger transitions in elements having low atomic numbers is much higher.

Kinetic Energies of Auger Electrons

AES includes measuring the K.E. of Auger electrons which are characteristic to the elements of the sample. Uncertainties of upto 1-2 eV are tolerable in measurement of K.E. for elemental identification. However, for analysing Auger chemical shifts, this uncertainty in K.E. should not exceed 0.1-0.2 eV. Nowadays, databases are available which can be compared with the obtained results however, earlier AES relied on measuring the K.E. of electrons. The steps involved can be summarized as follows:

K.E. of an Auger electron is expressed as the difference in energies of singly ionized initial state and doubly ionized final state. Since first principles calculations of this energy is highly complex, uncertainties upto ~10eV are acceptable in most situations, several approximations are to be made in order to determine the K.E. of Auger electrons. Suppose in a hypothetical ABC type Auger transition, a vacancy is created in A level by removing an electron from it; this vacancy gets filled by an electron from B level; and Auger electron is emitted from C level. Let the B.E. of electrons in neutral atom for A, B, and C levels are E_A, E_B, and E_C respectively. When electron from B occupies vacancy in A level, an energy $(E_A - E_B)$ is transferred to the third (C level) electron, which is the Auger electron. Thus, the K.E. (E_{ABC}) of Auger electron can be expressed as:

$$E_{ABC} \approx E_A - E_B - E_C$$

From Equation above, it is evident that the K.E. of Auger electron does not depend on the energy of the incident beam, rather it only depends upon the energy levels of the atoms present in the sample. Thus, the K.E. of Auger electrons represent the elemental composition at the surface of the sample. Each element shows distinct set of Auger peaks corresponding to its unique set of energy levels. For instance, the energy for KL_1L_3 Auger transition in aluminium can be calculated as:

$$\begin{aligned} E_{KL_1L_3} &\approx E_K - E_{L_1} - E_{L_3} \\ &\approx 1560 - 118 - 73 \\ &\approx 1369\,\text{eV} \end{aligned}$$

The measured value of this energy is 1354 eV. Two corrections can be considered to avoid the large disparity. Equation $E_{ABC} \approx E_A - E_B - E_C$ needs to be altered to include work function of spectrometer because Auger energies are usually referenced to the Fermi level. If good electrical contact exists between the sample and sample holder, then the Fermi levels of the sample and instrument are equal. Therefore, the K.E. of ABC Auger electron is approximately given as:

$$E_{ABC} \approx E_A - E_B - E_C - \phi_A$$

Here, φ_A represents the work function of spectrometer and its typical value is ~ 4 eV, E_{ABC} come out to be 1365 eV as the energy of $KL_1\,L_3$ Auger transition in aluminium. This reduces the disparity in measured and calculated values to 11 eV. For a neutral atom, Equation $E_{ABC} \approx E_A - E_B - E_C$ can be further filtered by considering the change in B.E. of a level accompanying ion formation:

$$E_{ABC} \approx E_A - E_B - E_C^* - \phi_A$$

Here, E_C^* represents the B.E. of a level in the presence of a core hole. Equation above can be employed to determine energy of Auger electrons based on various empirical approximations.

Instrumentation

Surface sensitivity in AES technique evolves from the fact that in solid materials, emitted electrons generally possess energies in the range of 50 eV - 3 keV and relatively shorter mean free path. Therefore, escape depth of electrons is limited to few nm in the target surface which gives AES a high surface sensitivity. Due to lower energy of Auger electrons, typical AES runs under the condition of ultra-high vacuum. This prevents electron scattering from residual gas atoms and development of a thin "gas (adsorbate) layer" on the sample surface which results in degradation of the analytical performance.

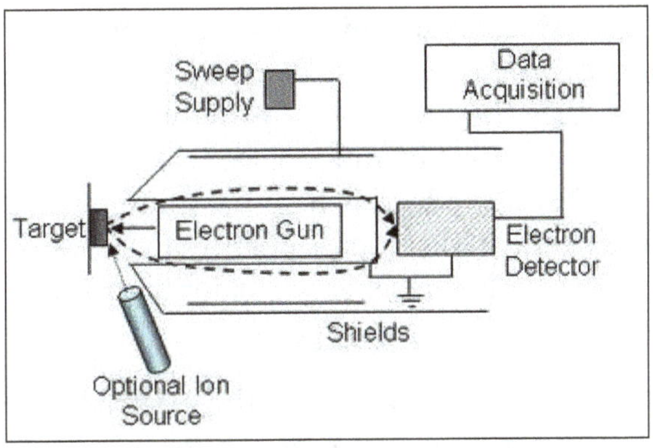

Schematics of AES experimental set up.

A standard AES setup involves the following components: (a) electron source and electron optical column forming an electron probe on the sample surface; (b) ion optical column to clean the surface and sputtering for depth profiling; (c) electron energy analyser; (d) a secondary electron detector and a pulse counter; and (e) computer control and display systems. Modules (a-d) are located inside an ultrahigh vacuum chamber.

- Electron Optical Column: An optical column focuses the electron beam (from an electron source) to the sample surface. The electron beam must be mono energetic, of small size, and highly bright. Electrons irradiating the sample have energy in 2-10 keV range with a beam current of 10^{-8} to 10^{-5} A. Scanning Auger microscopy uses energy upto 35 keV and current 10-9A to create a beam of ~100 Å diameter. The electron source must have prolonged life and higher temporal stability. Four major types of electron source are: (i) Tungsten thermionic emitter with ~2700 K operation temperature. They produce low current density. Owing to low brightness and large beam sizes, they are preferred for AES analysis where high lateral resolutions are required. (ii) The lanthanum hexaboride (LaB_6) thermionic emitter with operating temperature of ~1850 K and lower work function than tungsten, can provide high lateral resolutions since they generate high current densities.

(iii) Tungsten single crystal based cold field emitter, working at room temperature in the presence of a high electrostatic field and creates high brightness. However, this emitter becomes unstable on encountering residual gases and thus ultra-low pressures (10^{-10} torr) are needed. (iv) Hot field or ZrO_2/W Schottky-type field emitter, operable at ~1800 K. This type of emitter can produce high current density.

- Ion Optical Column: Ion source of electron impact type are used in AES equipments for depth profiling and sputter cleaning purposes. Heated filament generates electrons which are then accelerated by cylindrical grid such that they have enough energy to ionize gas atoms upon collision. The ions thus produced are accelerated into a focusing lens column. Typically, inert gases (e.g., Ar and Xe) are used with a hot tungsten filament.
- Electron Energy Analysers: The number of electrons emitted (N) are measured as a function of their energy (E) by using electron energy analysers. Most widely employed analysers are:(i) RFA or retarding field analyser. (ii) CMA or the cylindrical mirror analyser. (iii) CHA or concentric hemispherical analyser. Retarding field analysers suffer from inadequate energy resolution and poor signal-to-noise ratios due to average transmission efficiency. Thus, RFA is generally employed in low energy electron diffraction investigations, and not preferred for AES. CMA has high transmission efficiencies, compact size and is simple to use, making them suitable for AES. CHA is used for AES whenever chemical state analysis is performed since they have high resolutions. CHAs include an input lens and hemispherical analyser, and are a commonplace in XPS energy analysers. CMA and CHA are band pass filters (BPF) while RFA is high pass filter. The BPF allows passage of electrons within a band of energy (E) at a pass energy (E), leading to the energy resolution of E/E.As RFA detects electrons having energy greater than a specific energy, E, the spectrum is differentiated once to determine N(E) spectrum.
- Electron Detector: After exiting the analyser, the electrons reach the detector which amplifies and counts them using an electron multiplier (channeltron or microchannel plate, MCP). The interior of channeltron (cone-shaped dynode) is coated, such that when one electron strikes it, many secondary electrons are generated. These secondary electrons are then accelerated to the anode. Several intervening collisions cause an avalanche effect, leading to a gain of ~10^8 for a channeltron. MCP comprises several small channeltons joined together to create a disc. Electron intensity is measured via pulse counting.
- Computer and Display System: Sophisticated computer control, data processing, and display systems are common to all modern AES instruments. The functions of computer control system are: (i) setting up conditions for analysis; (ii) efficient collection and storage of data; (iii) data processing; and (iv) displaying results in the form of spectra.

In this experimental configuration, first the focused electrons get incident on a solid sample and then the emitted electrons gets deflected towards a cylindrical mirror analyser. In detector, Auger electrons gets multiplied and signal is advanced for data processing. Finally, the accumulated Auger electrons are recorded as a function of energy against broad secondary electron background spectrum. Because the Auger peaks have small intensities relative to the background noise, AES is generally operated in derivative mode that highlights Auger peaks by modulating the electron collection current by applying small AC voltage. Using setup shown in figure, by detecting the signal at frequency 'ω' and plotting one can get the Auger fine structure. In this Auger electron structure, one can clearly see the little secondary peaks neighbouring the main Auger peak. The secondary peaks arise because an element may be present on the surface in different chemical states.

Microscopy

Transmission Electron Microscopy

TEM is also known as conventional transmission electron microscopy or CTEM. Max Knoll and Ernst Ruska invented it in 1933 in Berlin. Recent electron microscopy (based on transmission) commonly contains a beam column which is around 2.5m tall and has a 30cmdiameter, and has an ability to attain a 2Å resolution. This technique is utilized for analysing the surface structure, i.e., morphology, surface imperfection, i.e., defects, crystal structure of the atom, size of the particle and also sample's composition.

Design of TEM is similar to a light microscope.

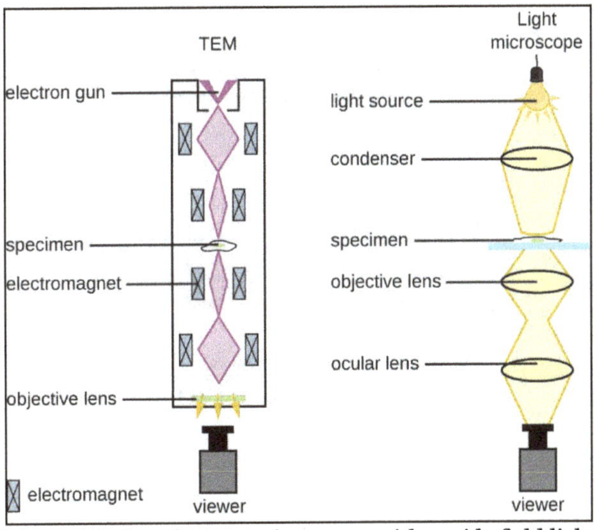

Similarity of a transmission electron microscope with a wide field light microscope.

Working Principle

An extremely thin sample is required for scanning in TEM from which electron beam is passed through rendering its interaction with the sample as a result of which image is produced. This image can be magnified and focused on the device used for imaging, like a fluorescent screen, on a photographic film layer, or to be identified by a sensor like a CCD camera.

Instrumentation of TEM:

- Source of electron.
- Gun based on Thermionic Emission.

- Beam of Electron.
- Electromagnetic lenses.
- Vacuum chamber.
- Two Condensers lenses, objective and intermediate lens.
- Sample holder and stage.
- (Imaging Device) Phosphor or fluorescent screen.
- Computer.

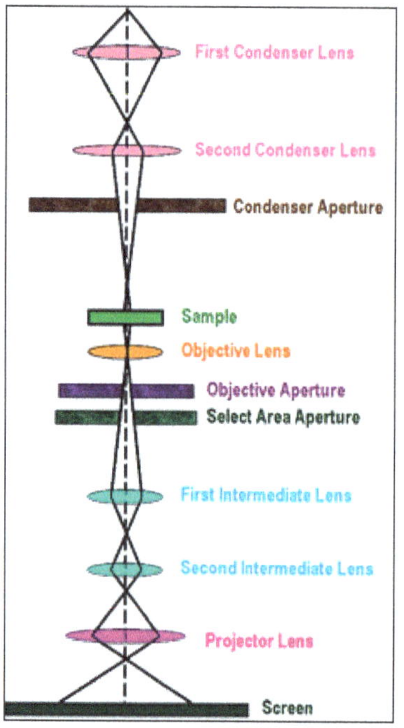

Instrumentation of TEM.

Electron Gun

Electrons can be produced either by thermionic emission or by a process called cold field emission. During thermionic emission, a very fine tip of a tungsten filament, a LaB6 crystal or a ZrO/W Schottky emitter, is heated by an electrical current flowing through it, thereby enabling the emission of electrons. The electrons leaving the filament have a low energy and, therefore, need to be accelerated to the desired speed before entering the electron column. A high voltage between the electron source (cathode) and an anode plate is applied leading to an electrostatic field through which the electrons are guided and accelerated. During cold field emission, the electrons can escape from an extremely fine tungsten tip without heating (at room temperature). The advantage of cold field emission sources is the very high yield of electrons and the very low chromatic aberration of the electrons allowing imaging at atomic resolutions. These instruments are very costly and require very high vacuum. The Electron gun working can be controlled based on 3 parameters:

- The accelerating voltage.

- Current of the filament (and therefore its temperature).
- And the Wehnelt cap bias voltage.

The temperature of the filament tip is controlled by the filament current which in turn controls the amount of emitted electrons. The filament current is increased till the number of emitted electrons no longer increases, which actually means that filament is saturated in order to maximize the emission.

The passing current between the system having high voltage and ground is controlled by bias resistor setting which in turn is controlled by the gun bias. When small bias voltage is used, the wehnelt negative potential is in effective in comparison to the filament, leading to poor focusing of the electrons that are accelerated toward the anode. This results in beam spreading, causing it to appear weak on the screen. When the biasing is increased, the focusing action is improved therefore the effective beam brightness is also increased; but, beyond a certain value the Wehnelt is so negative in comparison to the filament that the brightness starts to decrease because electrons are not permitted to emit from the filament or, in a case they are emitted, they are repelled back in the direction of the filament. The point at which the finest brightness of the beam is attained is determined by the distance between the Wehnelt and filament.

Electromagnetic Lenses

Electromagnetic lenses consist of a huge bundle of windings of insulated copper wire, a soft iron cast and pole piece. A magnetic field is induced by the current in the winding and reaches its main strength at the pole piece of the lens. The accelerated electrons entering the magnetic field are deviated by Lorentz forces. The direction of both magnetic field as well as electrons defines the resultant force which is always perpendicular to the plane. In conclusion, the electrons take a circular path through the lens system.

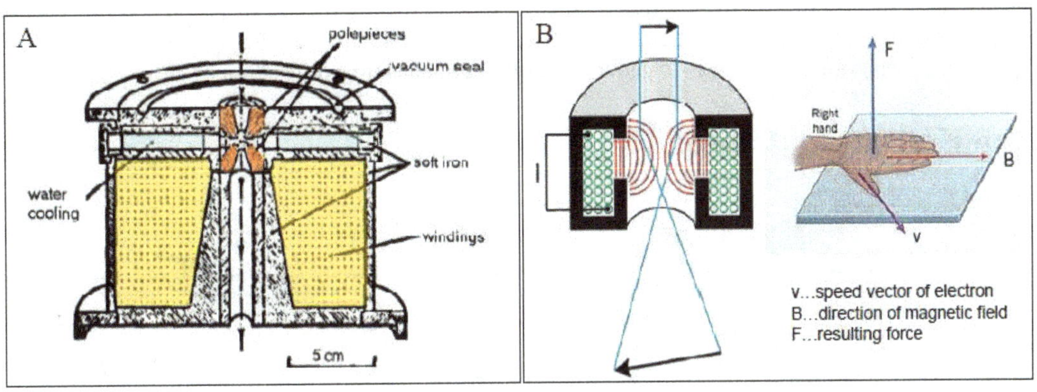

Electromagnetic lens. A) The magnetic field is strongest in the area of the pole piece; B) Electrons passing the magnetic field are deviated perpendicular to the plane defined by the magnetic field B and the velocity vector v.

Condenser Lens System

The beam diameter is reduced and controlled by condenser lens system. The purpose of the first condenser C1 lens (or spot size) which is a strong lens is to de-magnify the electron source image by around X1/100 to provide a small "point" source at the "crossover" that is more coherent than the large (50 μm diameter) tip of the filament. The purpose of the second condenser C2 lens (brightness or intensity) which is a weaker lens is to project the de-magnified image of the source on top of the sample by a magnification of X2, giving an overall demagnification of X1/50. Illumination spread onto the screen is controlled by this lens. A part named condenser aperture is positioned just below or sometimes between the condenser lenses; its role is to collimate (i.e. making parallel) the beam of the electron as well as modification in its intensity.

Objective and Intermediate Lenses

The reason behind the back focal plane being very close to the lens itself is because the magnification factor of the objective lens is larger. Aperture of the objective (it is the middle aperture on the column) is mounted in the back-focal plane. The selected area aperture sits in the first image plane below the specimen, which is below both the objective lens and the objective aperture. By altering the first projector lens excitation (also known as intermediate lens or diffraction lens), either an image or a diffraction pattern is produced.

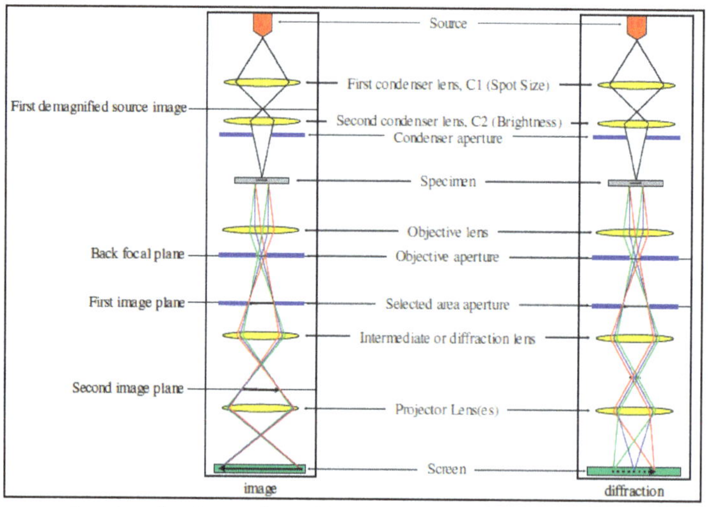

Diagram showing the formation of both an image and a diffraction pattern.

Specimen Holders and Stages

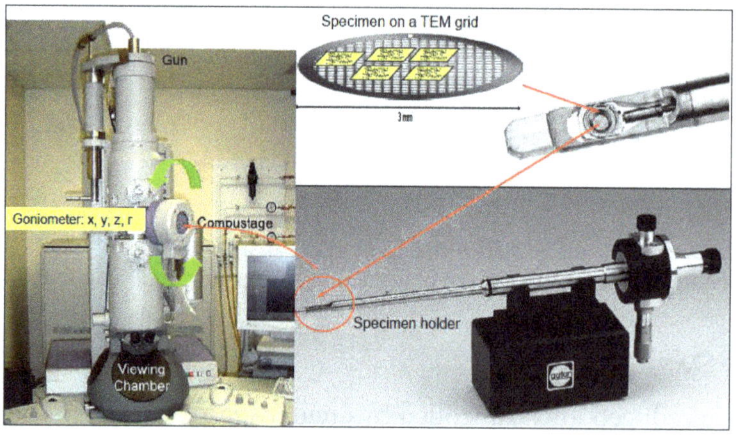

Thin sections of a specimen on a TEM grid, holder tip and complete specimen holder, which is introduced into the goniometer of the TEM through a vacuum lock.

In TEM, the electron column does not offer a lot of space for the sample. Further, the sample should be fine (thin) so that the electrons can penetrate the specimen to produce an image. The average thickness of a biological specimen should be around 70 nm for a TEM with an acceleration voltage for the electrons of ~100 kV (higher voltages allow the investigation of thicker samples). Thin sections of the sample are mounted on copper grids of 3 mm diameter, which are available in a wide variety of materials and mesh sizes. The grids with the sections on top are attached in a holder and introduced into the goniometer of the TEM through a vacuum lock, since the system always stays under high vacuum. The goniometer is the mechanical setup which enables highly precise and stable control of the specimen holder during imaging. Any drift or instability results in an blurred image, particularly at high magnifications.

Vacuum System

Vacuum system is employed in electron microscopes for 4 reasons:

- As electrons are readily scattered, electrons have a mean free path of ~1cm at atmospheric pressure; however, at 10^{-6} Pa they can have mean free path as high as 6.5m.
- The purpose of the vacuum system is to provide insulation between the filament of both anode and cathode as well as in the region around the field emitters, thus hampering undesirable discharge of the electron gun.
- In order to inhibit the oxidation and 'burning out' of the filament, oxygen is eliminated around the filament.
- Samples contamination is decreased by reducing the interaction amongst electron beam and molecules of the gas.

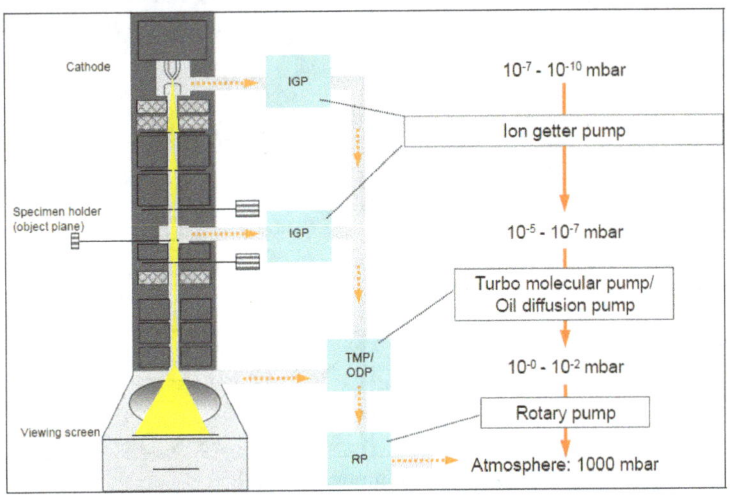

A typical vacuum system for a TEM. RP-rotary pump, TMP-turbo molecular pump, IGP ion getter pump, ODF-oil diffusion pump.

Various microscope parts are differently vacuumed as per their requirements. The gun requires 10^{-9} Pa vacuum, while the specimen requires 10^{-6} Pa and the projection chamber plus camera requires 10^{-5} Pa. Vacuums can be categorized as: rough (100 - 0.1 Pa), low (10^{-1} – 10^{-4} Pa), high (10^{-4} – 10^{-7} Pa), or ultrahigh (< 10^{-7} Pa).

Phosphor or fluorescent screen (Imaging Device): There are 2 procedures for specimen observation in TEM as shown in figure:

- Image mode
- Diffraction mode

In case of image mode, the electron beam hitting the sample is controlled by condenser lens and aperture, the beam which is transmitted will be focused and enlarged by objective and projector lenses, and the image is formed on the screen with identifiable information in relation to the microstructure of the sample. In case of diffraction mode, at the fluorescent screen a diffraction pattern (of electron) is obtained which originates from the electron beam illuminated sample region. The pattern of diffraction is completely similar to that of a pattern of X–ray diffraction. The spot pattern is produced by a single crystal on the screen whereas poly-crystal produces a pattern of powder or ring. The purpose of the image mode is to analyse microstructure, e.g. the grain size, and lattice defects, whereas the use of diffraction mode is to examine crystalline structure.

Image Modes of TEM

In TEM, the 2 primary image modes vary in the manner of using the objective aperture as filter in electron optics system. These modes are:

- Bright field microscopy
- Dark field microscopy

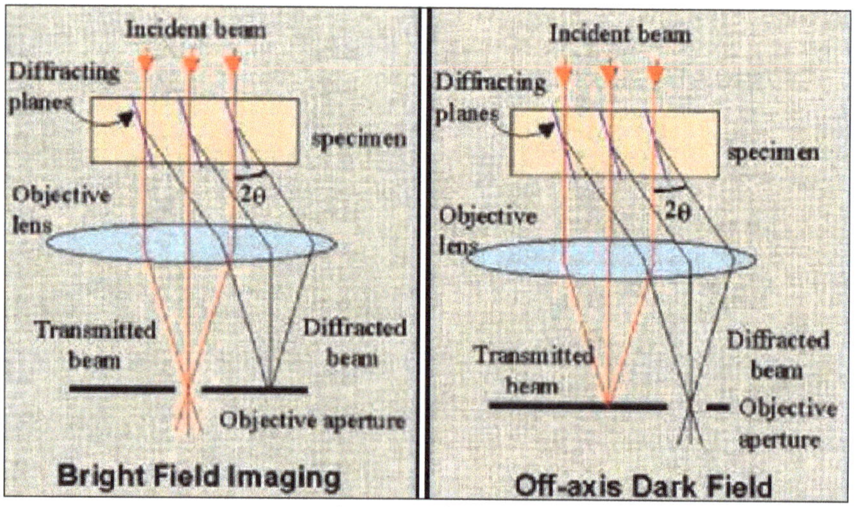

Two image modes of TEM.

In bright field imaging, the image of the sample is created by the electrons that pass through the film without diffracting. A diaphragm is used to stop the diffracted electrons. In the corresponding dark field imaging mode, the image is formed by the diffracted beam. The technique is called as bright Field which is mainly sensitive to extended crystal lattice defects in an otherwise ordered crystal, e.g., dislocations. The electron rays corresponding to bright field and dark field imaging are shown in Figure.

Electron Interaction with Matter

The interaction between the electron beam and the sample is coulombic. The negatively charged electrons can interact strongly with the electron cloud in the solid and also the positively charged nucleus. In contrast, X-rays are EM radiation and they only interact with the electron cloud. In TEM, for imaging purposes, only the forward scattered electrons are of interest. There are two main types of scattered radiation:

- Elastic: This represents coherent scattering (mainly) with no loss of energy. There is also a phase relation with the incident radiation.
- Inelastic: The energy of the scattered electrons is lower than the incident beam. These are also incoherent radiation with no phase relation with the incident radiation.

TEM Sample Preparation

Significant part of TEM is its sample preparation for the analysis. There are two main conditions for TEM sample preparation:

- Electron transparent sample must be used. If not the whole sample at least the ROI should be thin. The allowed thickness value for the metallic samples is 30 - 50 nm. Usually, 100 nm is an upper limit for the sample thickness.
- The sample must be mechanically strong for treatment.

TEM samples are either self-supported or mounted on a grid for analysis. Copper grids are the most commonly used, though for high temperature work Mo grids are used. For nanoparticles and thin films aC film is used as support. A-C has low contrast in the TEM and will not obscure the contrast arising from the specimen. Some typical TEM grids are shown in figure.

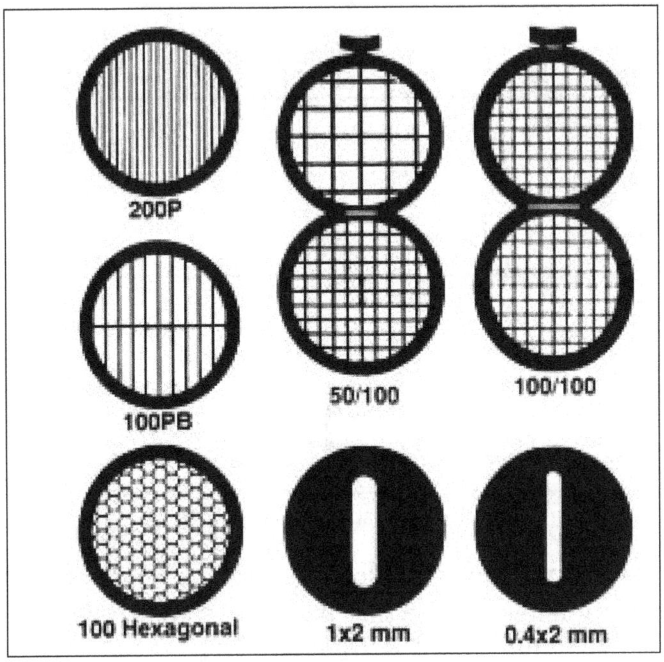

Typical TEM grids.

Thinning the Sample by Different Techniques

Electrolytic Polishing

Electrolytic polishing is used for conducting samples like metals/alloys in order to produce samples that are electron transparent. The initial sheet thickness can be around a few hundred μm. This can be prepared by rolling or grinding bulk specimens. Similarly, metal coatings on substrates can be peeled off and used for the final thinning. Thin discs can also be cut from bulk specimens. This process is called coring. These discs are thinned by electrolytic polishing. Electrolytic polishing technique is the window technique. The sample is made the anode and a thin stainless sheet is made the cathode. The sample is immersed in the electrolyte, which is usually cooled by water or liquid nitrogen. Perchloric acid is usually used as the electrolyte. The sample edges are covered by lacquer to expose a 'window', hence the name. The experimental setup and the hole generation are shown in Figure.

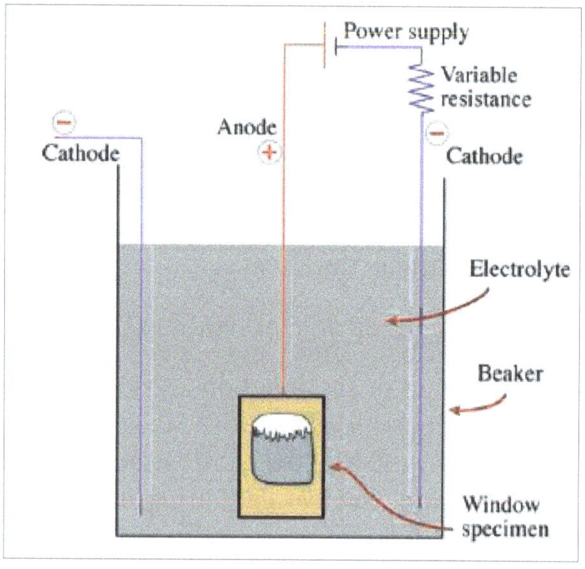

Window polishing technique.

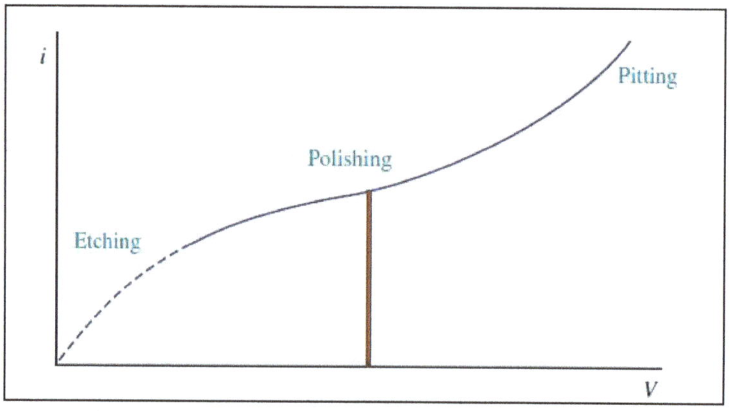

I - V characteristics during polishing.

When a current is applied, the material is dissolved from the anode (sample) and deposits on the cathode. The rate of dissolution depends on the current and applied voltage. The I - V characteristics are shown in Figure. Depending on the current and voltage, there are three regimes - etching, polishing, and pitting. The edges are coated so that material removal will start within the window. Once a hole is formed within the window, the sample is pulled out. The region around the hole is usually electron transparent and can be mounted on a TEM grid.

Ion Milling Technique

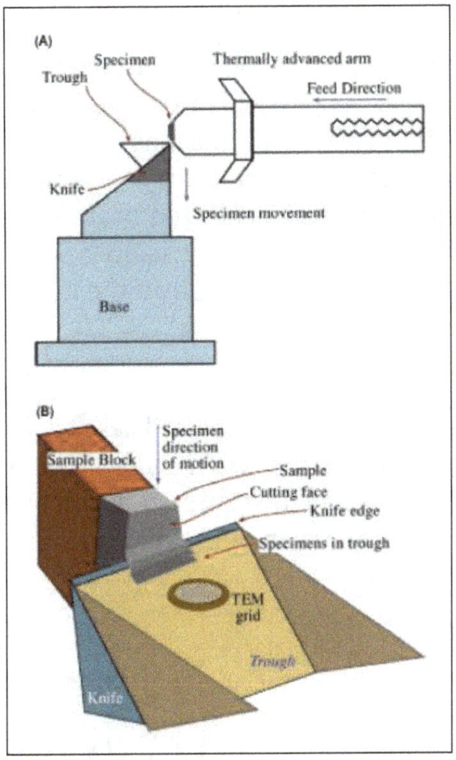

Schematic of ultramicrotome technique.

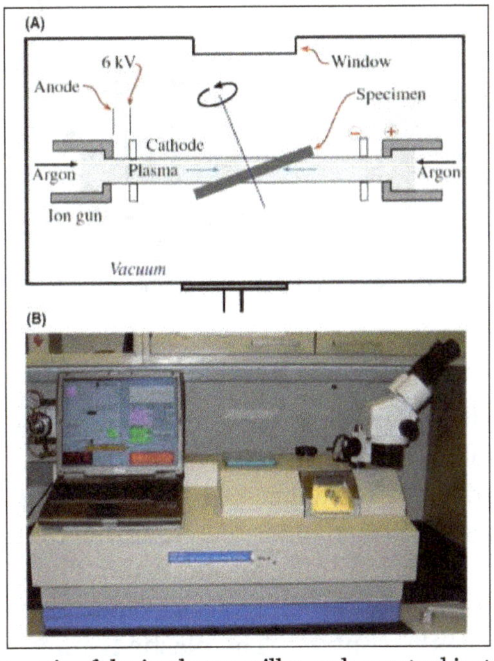

The schematic of the ion beam miller and an actual instrument.

For non-conducting samples, usually, grinding and polishing steps are used in order to reduce sample thickness. Sometimes, an ultramicrotome is used in order to generate thin samples. These can be either electron transparent or can be used as the starting material for further thinning. The schematic of the technique is shown in Figure. For samples, where ultramicrotome cannot be used then a standard tripod polisher is used in order to thin the sample. This produces samples that are a few nm thick. The final polishing step is done by an ion beam miller.

The sample is bombarded with high energy ions or neutral atoms. Usually, Ar ions are used and they are formed by passing the Ar gas though a high voltage (4 - 6 keV). The sample is held in vacuum and also usually cooled by liquid nitrogen. The ions are incident on the sample to sputter the material away. To minimize ion penetration the beam is usually incident at a low angle (≈ 20°), if the angle is very small the sputter rate is small. Ion beam is highly controlled and a localized process but it is time consuming. Sputter rates are usually a few Åper second so that creating an electron transparent sample can take hours, especially if the initial thickness is high.

Cross Section Sample Preparation

Slices from the sample are cut using a diamond slicer. These slices are placed between spacer layers and then glued on to a grid. The slices are glued in such a way that the interface is parallel to the slot in the grid. This sample is then thinned by standard tripod polishing until it is a few μm thick. The final sample is thinned using an ion beam miller to create an electron transparent sample.

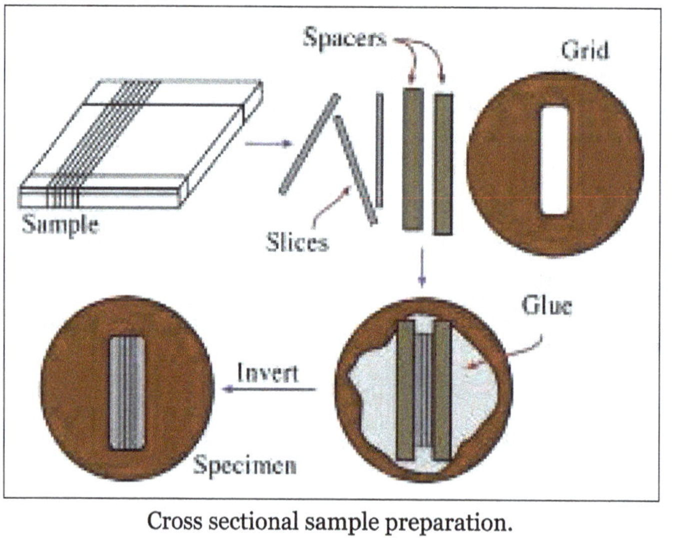

Cross sectional sample preparation.

Replica Technique

Replica technique is used for studying bulk specimens which cannot be destroyed to prepare electron specimens. It is also useful for studying surface topography features and precipitates though SEM techniques have gradually replaced replica sample preparation. A replica of the sample surface is prepared using a plastic mold. The mold is then removed from the surface and the surface of the specimen is replicated by the surface of the plastic. A thin film of carbon or metal like Cr, Pt is evaporated on the surface of the plastic. Sometimes the evaporation is done from an oblique angle, shadow evaporation, to enhance the contrast. The plastic is removed by dissolving in a suitable solvent and the film is then floated on to a grid for analysis.

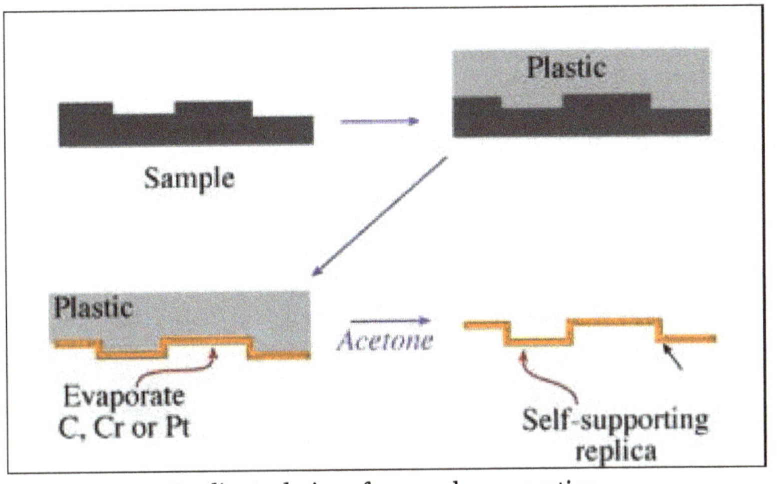

Replica technique for sample preparation.

Electron Probe Micro Analyser

The Electron Probe Micro Analyser (hereinafter, "EPMA") is an instrument to analyse which elements compose a substance, by irradiating electron beams onto the substance surface and measuring the characteristic X-ray that is generated.

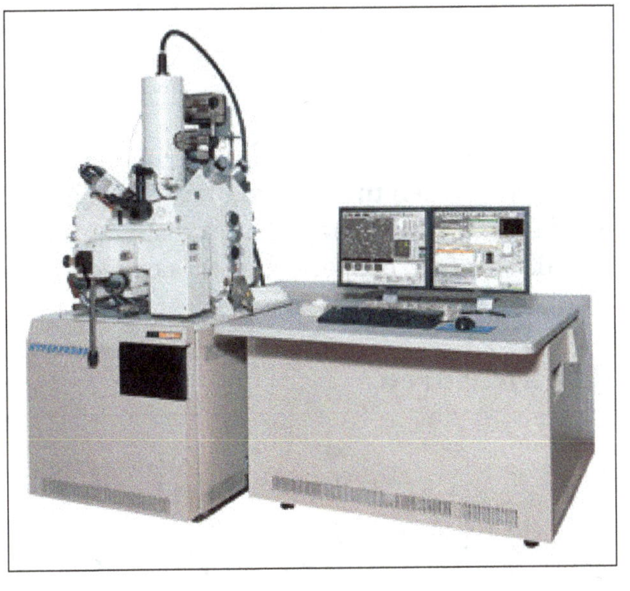

EPMA was devised and presented by Dr. Castaing of France in his academic dissertation in 1951. EPMA today is playing an active role worldwide, as an all-purpose, multi-functional tool equipped with electron optics technology, X-ray spectrometry technology, system control and data processing technology. EPMA has evolved into an instrument that can handle elemental analysis of sub-micron areas as well as observation, analysis, and image analysis for areas as large as 10cm square.

This instrument is used for basic research in a diverse range of fields, such as steel, minerals, semiconductors, ceramics, textiles, medical and dental materials, medicine, and biology, as well as for application research and product quality management.

Mechanism of the Instrument

What is the Structure of EPMA?

The figure shows the structure of an EPMA device.

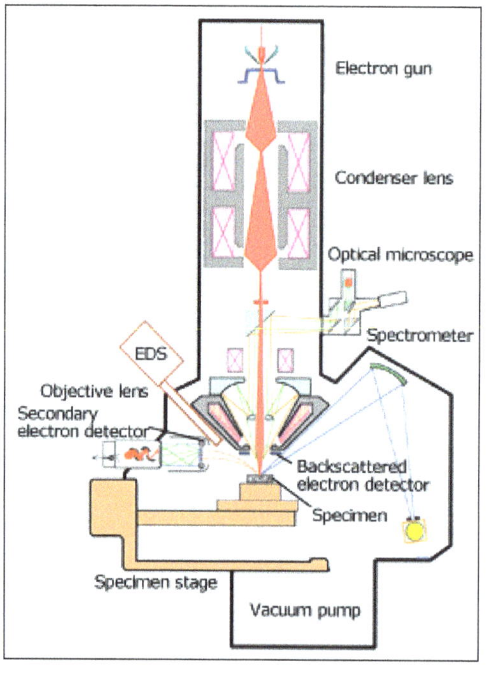

Electrons emitted from the electron source are accelerated at a certain accelerating voltage and collimated through electron lenses. When this electron beam hits a specimen, X-rays are generated from the specimen. By dispersing this X-ray using a dispersive element, the composition of the specimen can be examined. This type of spectrometer is called a wavelength dispersive X-ray spectrometer (WDS).

What Signals Should be Measured?

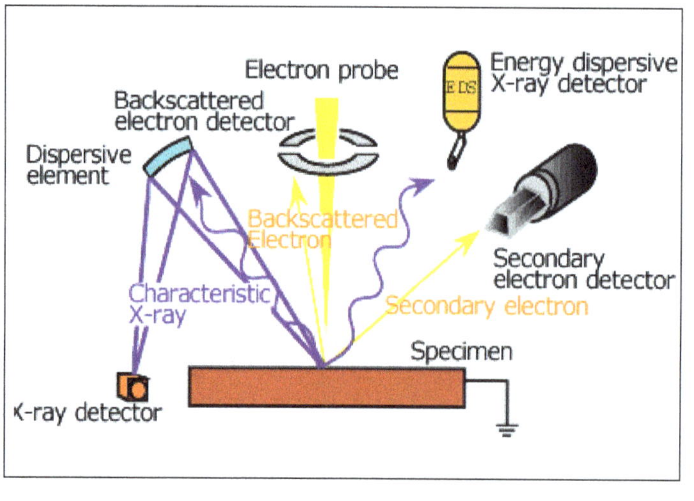

When accelerated electrons hit a specimen, in addition to the X-rays, particles and electromagnetic waves carrying various kinds of information are emitted. With EPMA, signals such as the characteristic-X-rays, secondary electrons, backscattered electrons, etc. are detected by the appropriate detectors and that information is utilized to find the area of interest on a specimen, and for analysis.

What is the Difference Between WDS and EDS Spectrometry?

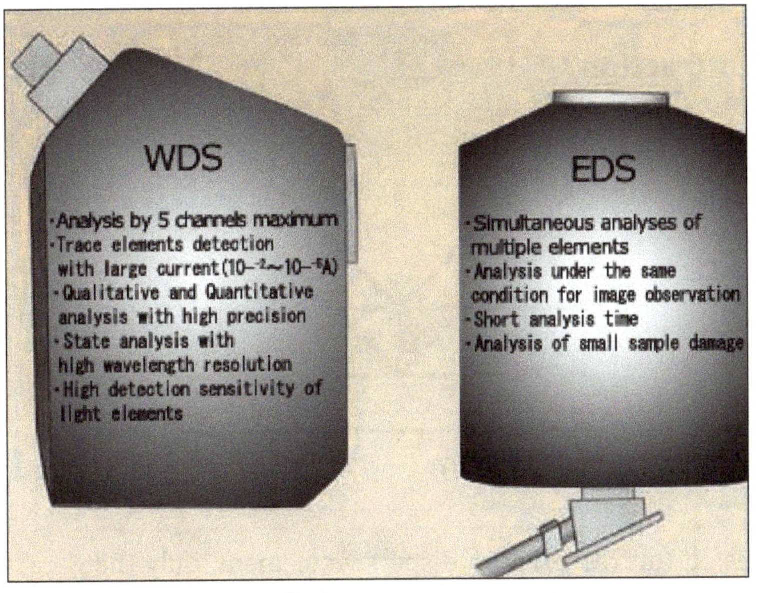

In addition to WDS, another spectrometry method for analysing the characteristic X-rays is energy dispersive X-ray spectrometry (EDS). There are differences between WDS and EDS due to the differences in the spectrometry method, as shown in the figure at the left.

X-ray Generation and the Detection Principle

What is the Mechanism of Characteristic X-ray Generation?

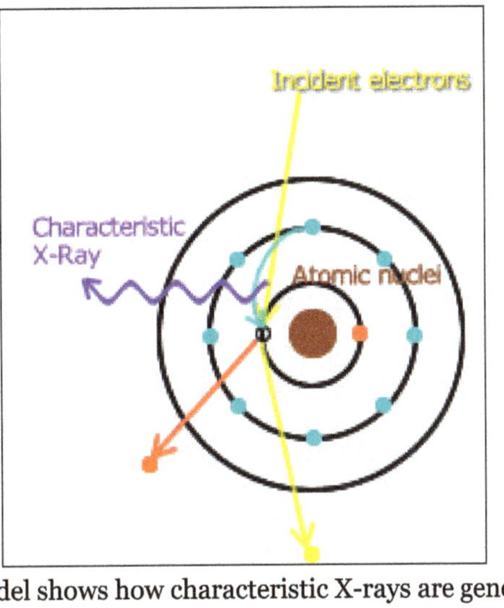

The model shows how characteristic X-rays are generated.

Accelerated electrons kick out an electron around an atomic nucleus. An electron from an outer shell at a higher energy level comes down to take the place of the ejected electron. This movement between energy levels is accompanied by the emission of an X-ray with a wavelength that is equivalent to the energy difference.

Since the energy level of an element is unique to the element, by investigating the wavelength of the X-ray emitted, or the energy, it is possible to identify which elements compose the substance being examined.

What is Bragg Diffraction?

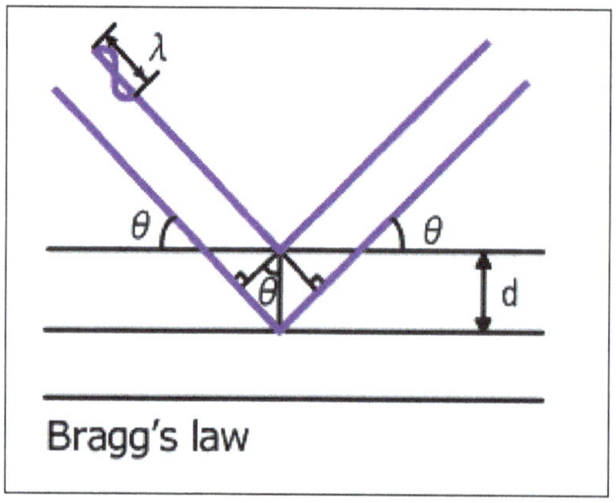

When X-ray is reflected (diffracted) by a dispersive element, only the X-ray which has a wavelength satisfying "Bragg's law $n\lambda = 2d\sin\theta$" is reflected. Wavelength dispersive X-ray Spectrometer (WDS) measures the wavelengths of X-rays by making use of this characteristic.

What is a Wavelength Dispersive X-ray Spectrometer?

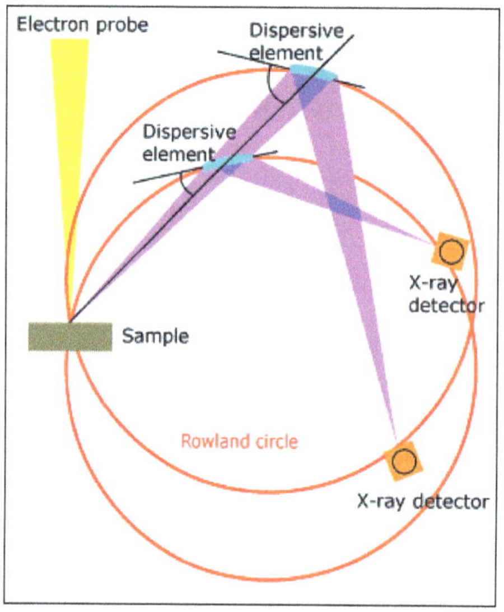

Wave-length dispersive X-ray spectrometer (WDS).

A wavelength dispersive X-ray spectrometer is constructed so that the specimen surface, analysing crystal and x-ray detector are positioned on a circle called the Rowland Circle, in order to satisfy

Bragg's law. Since a single analysing crystal cannot be used to analyse all elements, normally, there are several diffraction elements incorporated into one instrument.

Analysis Example of EPMA

What is Image Observation (Secondary Electron Image and Backscattered Electron Image)?

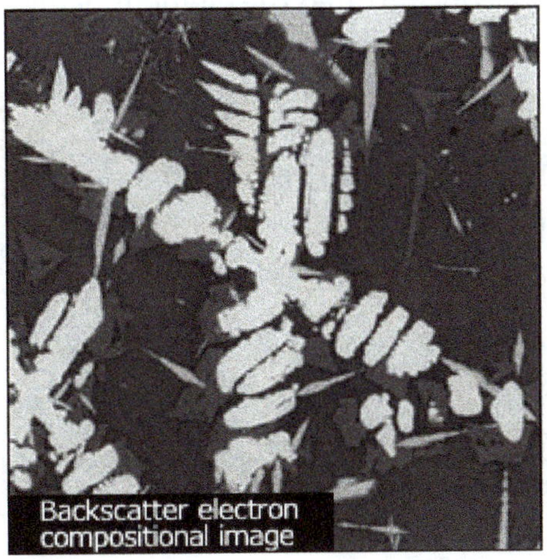

The first step of analysis is to identify the area of interest. Secondary electron images and backscattered electron images are very useful for this. Especially, a backscattered electron image can reveal a rough compositional distribution, so it is very convenient for EPMA analysis. In a backscattered electron composite image, an area with a large amount of signal is bright, indicating that the area is composed of an element with a larger atomic number than the elements in the darker areas.

What can we Learn from the Qualitative Analysis?

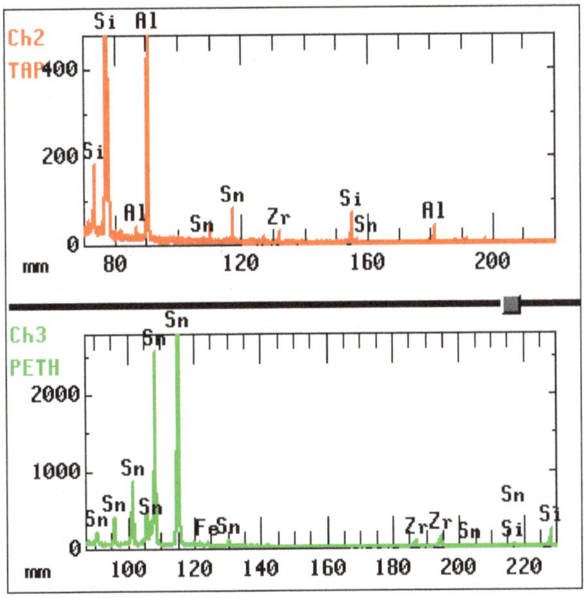

This example is the result of a qualitative analysis of a ceramic material (what elements are present). From the detected X-ray wavelengths it is possible to clarify which elements are present. In this case, we can see that Al, Si, Fe, Zr, and Sn are contained.

What is Area Analysis?

This shows the result of area analysis (where is each element distributed?) for Sn, and Si, which were detected in the preceding qualitative analysis. In area analysis, the more concentrated the red color on the color bar is, the more concentrated the element being analysed is. It is clear that there is a difference in distribution concentration between Sn and Si.

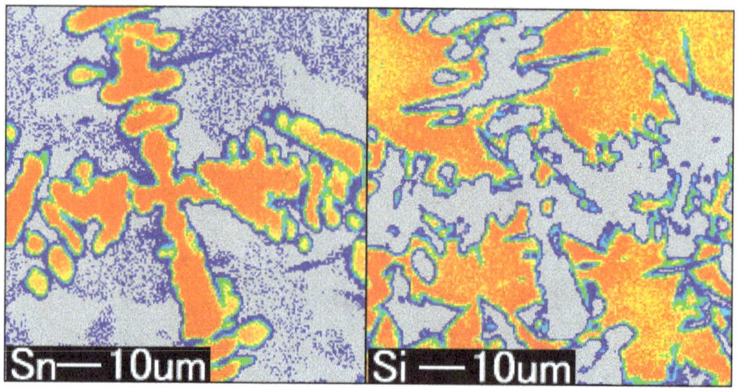

Energy Dispersive Analysis of X-rays

Also known as energy dispersive spectroscopy, and can be abbreviated as EDS, EDX, EDXS or XEDS sometimes it is termed energy dispersive X-ray microanalysis(EDXMA) or energy dispersive X-ray analysis (EDXA). It's an investigative method which is widely employed in elemental or chemical characterization of objects. The sample may be of any form such as solid thin films, solid powder, liquid sample or even a pellet etc.

Basic Analysis Principle of EDS

A typical set up of the EDAX is similar with that of SEM set up. More specifically, EDAX is actually inbuilt within the SEM set up. The working of EDX is based on electron focusing same as in SEM. Electron beam excitation is employed in SEM and scanning transmission electron microscopes (STEM) whereas the X-ray beam excitation is present in X-ray fluorescence (XRF) spectrometers. A detector converts X-rays into voltage signals; which feed a pulse processor, to measure the signals and pass them to data analyser for display and investigation. Si(Li) detector cooled at cryogenic temperatures by liquid nitrogen is often emploed; additionally, some modern systems come with silicon drift detectors (SDD) employing Peltier cooling systems.

To produce characteristic X-rays from an object, it is bombarded by either a highly energetic beam of charge carriers (electrons or protons) or X-rays. The atoms of the sample contain ground state (unexcited) electrons in discrete energy levels or electron shells bound to the nucleus. An electron from an inner shell may be excited by an incident beam, thereby removing it from its shell and generating a hole where electron was present before excitation. This hole can be occupied by an electron of a higher-energy shell. The difference in energies of the higher- and lower- energy shells is emitted as an X-ray. Quantitative measurement of the energy and number of these X-rays can be done with energy-dispersive analysis of X-rays. Since the energy of these X-rays is characteristic feature of energy difference between two shells and atomic structure of discharging element, EDAX can be employed to identify the elemental composition of an object.

Advantage of EDS is broad range of elements, i.e., from Boron to Uranium can be examined synchronously.

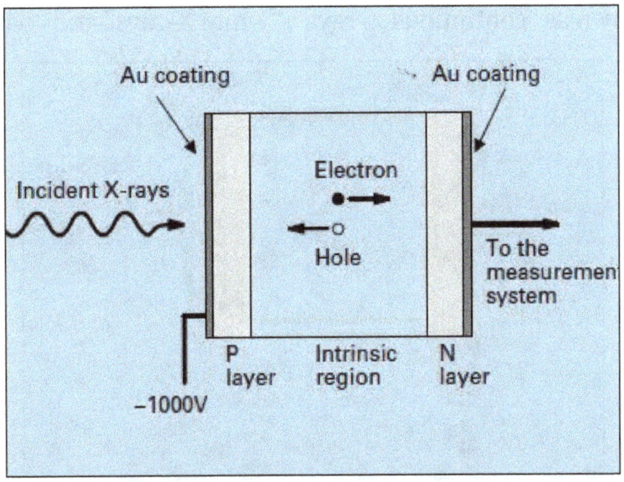

EDS semiconductor detector assembly.

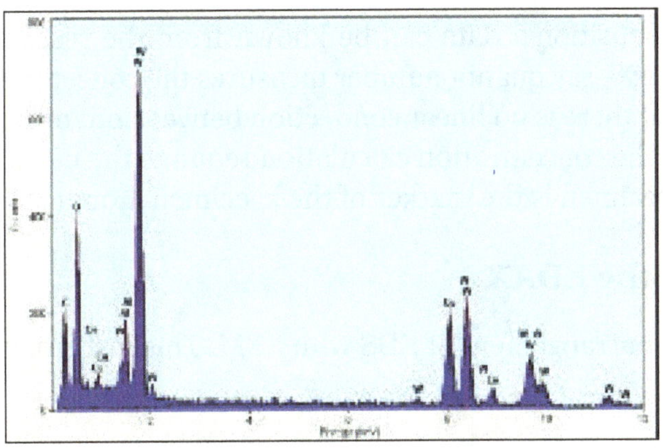

Example of EDS spectrum.

In Figure, X-axis represents the X-rays energy and the Y-axis depicts the X-ray counts. It is a valuable tool for Qualitative as well as Quantitative analysis. The above figure showing unique peak positions of various elements corresponds to the feasible transitions in its principal energy level.

Characteristic X-Rays Generation

When a sample is irradiated with highly energetic beam of electrons then a characteristic X-ray is generated. Due to irradiation when an electron is excited from the inner shell of its parent atom, the vacant place is occupied by the higher energy electron from the outer shell and the energy difference between the two shells is emitted as an X-ray. The joule of energy liberated in electron transfer is dependent upon from which shell it is transferring to which shell. Emitted X-rays energy as well as number through specimen is obtainable by an energy dispersive spectrometer.

These emitted X-rays are known as characteristic X-rays because their energies (wavelength) are specific to the element from which it is emitted. Thus, this technique can be used for elemental analysis. When K shell electrons are excited, they emit characteristic X-rays which are termed as "K Lines" and those emitted from L and M shells are called "L Lines" and "M Lines" respectively. When the element is heavy then its characteristics X-rays energies are also higher, therefore incident electrons of higher energy are required. Different types of X-rays are emitted when incident electrons strike the atomic nucleus and these are known as "continuous X-rays", "white X-rays" and "background X-rays".

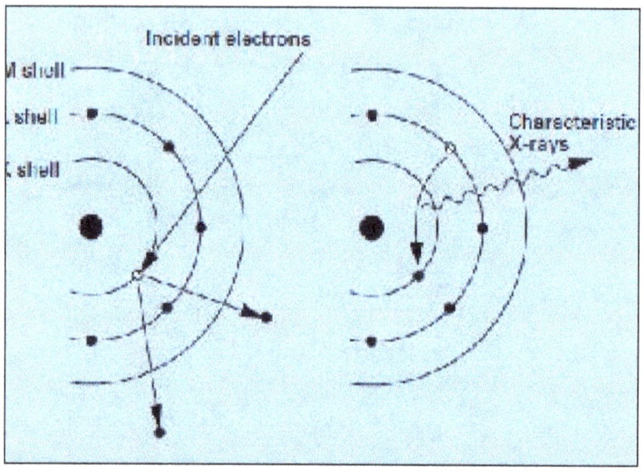

Principle of characteristic X-Rays Generation.

Sample's qualitative compositional data can be known from the respective peak positions with appropriate energies. The X–ray quanta number measures the concentration of the elements, i.e., the height of the peak, but there is no linear connection between quantum numbers and elemental concentration portions. The concentration calculation requires the net count rates and characteristic X-ray energy of a specimen is the marker of the specimen atomic number.

Instrumentation of the EDAX

Figure shows the working arrangement of EDS with SEM. This instrument is mainly composed of four components:

- Electron beam source
- X–ray Detector
- Pulse Processor
- Analyser

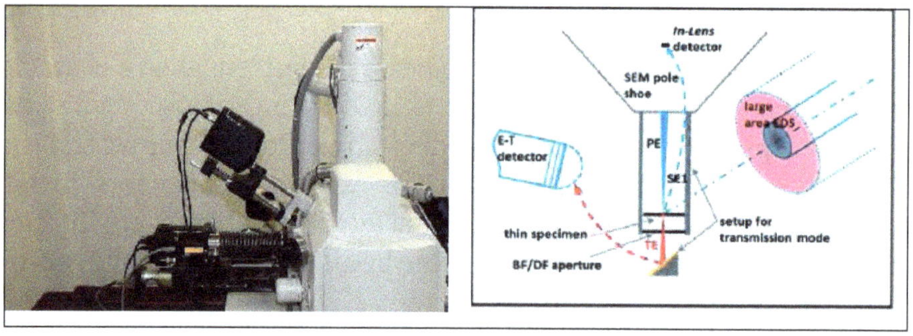

Working arrangement of EDAX (Energy dispersive analysis of X–rays).

Brief Description of Components

- Electron beam source: As EDS is employed with the scanning electron microscope therefore the same electron gun (Field Emission) is used as the incident electron source and to focus the beam lenses are used along with aperture. The energy of the electron beam has to be carefully selected to overcome the compromise between the resolution requirements and the production efficiency of X-rays. The produced X-rays are detected by two crystal spectrometers.
- X–ray Detector: The X-rays counts (which is the abundance of emitted X-rays) versus X-rays energies are measured by the EDS detector. Detector, a solid state device which is based on lithium drifted silicon. As X–rays strike the surface of the detector, a charge pulse is created. This charge pulse is directly proportional to the energy of the incident X–ray.

Parts of Detector

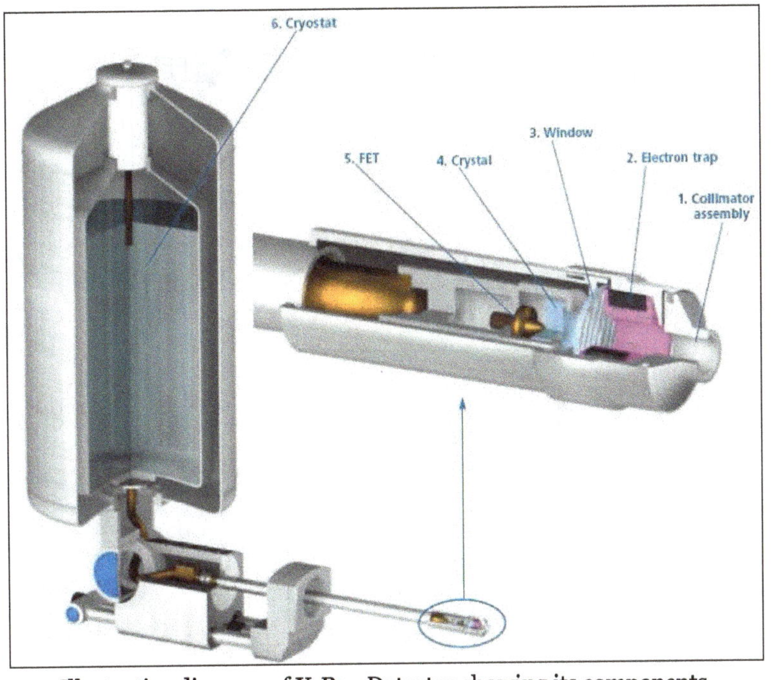

Illustrative diagram of X-Ray Detector showing its components.

- Collimator assembly: The purpose of the collimator is to provide restrictive aperture through which X-rays should pass in order to make a way to detector. The collimator ensures that only X-rays from the region being excited by the beam of electron are detected.
- Electron trap: Background artefacts are created when electrons penetrate the detector. The purpose of the electron trap is to guarantee that only X-rays should enter the detector. It comprises a pair of permanent magnets to strongly deflect the passing electrons.
- Window: The purpose of the window is to provide isolation from the microscope chamber to maintain vacuum inside the detector, at the same time being as transparent as possible for allowing low energy X-rays. Char-Fac comprises a thin window composed of polymer and transmits X-rays from elements which are heavier in comparison to Beryllium.

- Crystal: In detector crystal material used is silicon (Si), drifted with lithium (Li) [around 3mm thickness] for small levels of impurity compensation. When X-rays are incident on the crystal of detector then its energy gets absorbed by a series of ionizations within the semiconductor to generate electron-hole pairs. The absorption of every 3.76eV of X-ray radiation generates an electron-hole pair. Therefore, 1,966 electrons are generated by a Ni Kα X-ray photon (7,471 eV). The electrons promoted to the conduction band can freely move in the crystal lattice. Every excited electron leaves behind a 'hole' in the valence band, which acts as a free (positive) charge carrier within the crystal. When the front and back face of the crystal are biased with high voltage, these electrons and holes are swept towards the opposite poles, there by generating a charge signal, whose size is proportional to the incident X-ray energy.

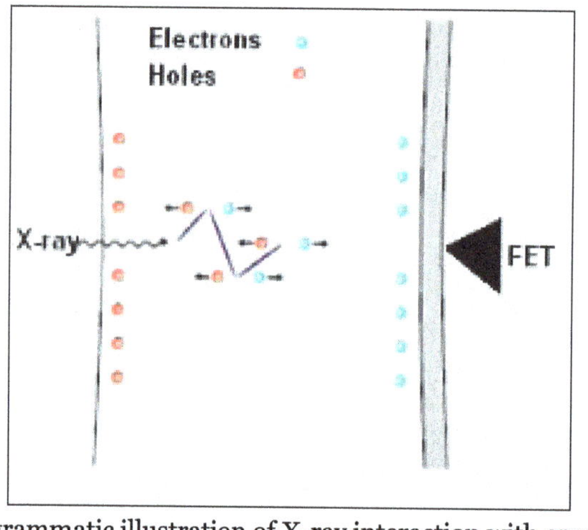

Diagrammatic illustration of X-ray interaction with crystal.

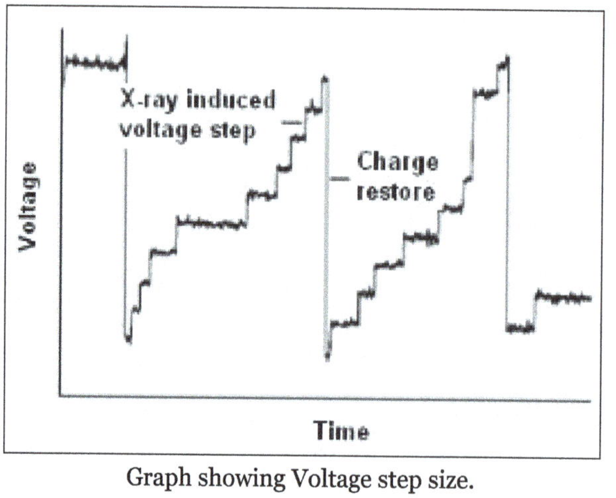

Graph showing Voltage step size.

- FET: The FET (Field Effect Transistor) performs first stage amplification which measures the charge released within the crystal by the incident X-ray and converts it to a voltage output. FET is nowadays integrated with the chips in order to greatly enhance the resolution along with the throughput which is achieved by reducing noise through minimizing the capacitance between FET and anode. During operation, charge is accumulated by each incident X-ray on the feedback capacitor, thereby producing sharp voltage steps. The step size of this voltage is proportional to the energy of the incident X-ray. Restoration of the build-up charge is required periodically to avoid preamplifier saturation.

- Cryostat: EDS detector is a self-enclosed vacuum system (called cryostat) with liq. N2, or a cryogenic pumping. The crystal and FET are mounted on cold finger within cryostat to reduce the noise by cooling. The cooling is necessary as the detector produces small charge pulses. At low voltages, there solution of the detector depends upon the noise. The X-ray peaks are typically 2-10 eV wide (FWHM). However, due to noise, this width can increase by at least an order of magnitude. The vacuum level should be maintained such that the condensation of molecules on crystal can be avoided.
    - Pulse Processor: A charge–sensitive preamplifier is employed to convert the charge pulse to a voltage pulse.
    - Analyser: Multi-channel analyser is used to sort pulses by voltage in the signals which are received by the analyser. The energy of the X-ray can be obtained by measuring the voltage of the charge pulses. This energy is then sent for display and data processing. Here, data is displayed as histogram of intensity vs voltage.

Table: Features of EDS.

Features	EDS
Measurable element range	B ~ U
Measurement method	Energy dispersive method with a Si (Li) semiconductor detector
Resolution	E = 130 ~ 140eV
Measuring speed	Fast
Multi-element simultaneous measurement	Possible
Damage/contamination of specimen	Little
Detection limit	1500 ~ 2000 ppm
Detectable X-rays per current	Many

Qualitative Analysis

X-rays generated EDS spectrum helps to carry out qualitative analysis to identify whatever elements are present in the sample when it is irradiated with high energy beam of electron. There are 3 methods of examination:

- Point Analysis: Spectrum is obtained from a point which is radiated with electron probe.
- Line Analysis: Spectrum is obtained from a specified line displaying one–dimensional desired elemental distribution.
- Mapping: Spectrum is obtained from a specified area displaying two-dimensional desired elemental distribution. It is often called as “Area Analysis”.

X-Ray Mapping

It is used to study particular elemental distribution. For analysis purpose, electron beam is scanned over a particular area and the related characteristics X-rays are acquired. If the peak to background ratio is very low then X-ray Map does not show the distribution of desired element, i.e., it shows the distribution of continuous X-rays. Sometimes X-Ray Map may not show the distribution of elements which are not of interest because characteristics X-Rays of elements of interest are close to those of not desired. It happens when energy difference between the desired element and undesired element is equal to the spectrometer energy resolution. It takes more time to acquire one X-ray picture as intensities of X-rays are less in comparison to that of secondary and backscattered electrons. The resolution of X-Ray Mapping is limited by "Area Analysis". Figure shows an example of X-Ray Mapping.

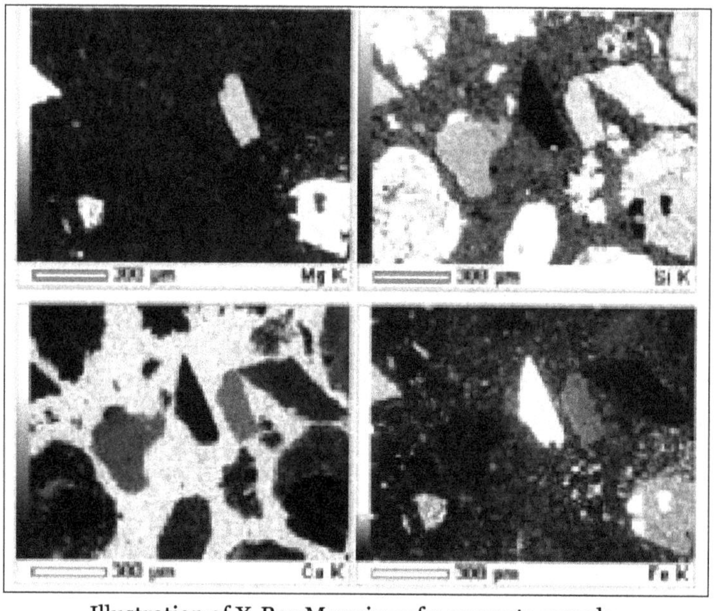

Illustration of X-Ray Mapping of a concrete sample.

The procedure which quantitatively evaluates the points upon the sample gradually while scanning the electron beam is called "Quantitative mapping". Precise elemental distribution can be acquired using "quantitative analysis" method even if Peak to Background ratio is low. It is a great advantage of this method.

Analysis Area

Incident electrons when strike the specimen surface, get diffused and lose their energy by generating characteristic X-rays. This X-ray generation is spread over an area of around a few micrometers under standard operating condition. Thus, a larger area needs to be analysed in order to examine the particle of the order of few nm range in a SEM image. The analysis area can be decreased by reducing the accelerating voltage. However, accelerating voltage cannot be reduced below a certain level, since the generation of X-rays requires high accelerating voltages. Another strategy to reduce the analysis area is by reducing the thickness of the sample under investigation.

Quantitative Analysis

As characteristic X-rays intensities are proportional to the particular element's concentration, therefore quantitative analysis can be performed. The concentrations of unknown elements in a sample can be obtained by matching the intensities of characteristic X-ray of standard sample with that of the unknown sample. However, X-Ray generated in the sample may be either absorbed within the sample or excite the X-rays from other elements before being emitted in vacuum. Therefore, correction is required in this analysis method for which the following requirements are prerequisite:

- Flat specimen surface.
- Electron probe should enter the sample in perpendicular direction.
- Distribution of elements in the X-ray generation area should be uniform.

There are always some appreciable errors in this analysis method because some specimens do not meet these requirements.

Non-Conductive Specimen Analysis

For non-conducting sample, metal coating is required in order to prevent the charging but if requirement is to detect light element then it is suggested to use thin coating film, as heavy metal coating can block the X-ray emission through the sample because the X-ray intensities are less in comparison to secondary and backscattered electrons. Analysing sample with low incident accelerating voltage can cause some problems like:

- High excitation energy based characteristics X-rays cannot be detected.
- Quantitative analysis accuracy is degraded.
- While performing "line analysis" or "X-ray Mapping" position shift in incident electron probe occurs.

Suggestion for such problem: Low vacuum SEM can be used to analyse non-conducting sample because residual gas molecules present in the sample chamber scatter the high energy electron beam thereby increasing the analysis area.

References

- Thermal-analysis-applications: instrument-specialists.com, Retrieved 25, June 2020
- Differential-scanning-calorimetry-dsc: photometrics.net, Retrieved 30, January 2020
- A-X-ray-Photoelectron-Spectroscopy, A-Physical-Methods-in-Chemistry-and-Nano-Science-(Barron), Analytical-Chemistry: libretexts.org, Retrieved 16, April 2020

3

Materials Properties: Physical, Mechanical and Chemical

All materials have specific and distinguishing properties. Physical properties are measurable like density, melting point, conductivity etc. Mechanical properties are the effects of different forces on the performance of a material. Various physical, electrical, biological, mechanical, thermal and chemical properties of materials are discussed in this chapter in detail.

Physical Properties

A material undergoes transition under the influence of temperature and pressure, and these changes are physical in nature, because their molecules remain intact.

Effects of Temperature on Substances

When temperature rises, a typical substance changes from solid to liquid and then to vapor, at a constant pressure. Some substance has several crystal forms in the solid state. The glassy state is also considered a solid. Transitions from one solid to another solid form, from solid to liquid, from liquid to vapor, from vapor to solid etc. are called phase transitions.

Phase transitions from solid to liquid, and from liquid to vapor absorb heat. The temperature of a system usually does not change as long as two phases are present. The (phase) transition temperature from solid to liquid is called the melting point whereas the temperature at which the vapor pressure of a liquid equals 1 atm (101.3 kPa) is called the boiling point. Thus, the measured boiling point depends on the atmosphere pressure. For compounds that decompose at high temperature, boiling point can be either specified at lower pressure or be replaced by the decomposition temperature. Thus, conditions as well as the value of boiling point listed in literature must be taken into account for application considerations. Boiling points of mixtures change with composition.

However, a glassy material becomes soft in a wide range of temperatures. The temperature at which the material becomes soft (behave molten like) is called glassy temperature, but it may be a range of temperatures. Behavior of a substance as the temperature changes must be carefully considered in its applications. Behavior of a mixture as temperature rises is different from its components. There is no theoretical way to predict the behavior of a mixture from its components, even if its exact composition is known. Addition of one or more materials usually changes the melting or glassy temperature of a substance. Thus, we often employ a blend (mixture) of materials whose behavior is acceptable within the desirable range of temperatures. Antifreeze for automobile radiator and deicing liquid for airplanes are examples of this application.

The behavior of mixtures as temperature and pressure change often requires a phase diagram to represent, and there are several models of two-component systems. A phase diagram of a many-component system requires a lengthy study.

- Phase Transition at Constant Pressure:
    - Temperature
- Vapor:
    - Boiling point
    - Heat of vaporization
- Liquid:
    - Melting point
    - Heat of fussion
- Solid

Combined Effect of Temperature and Pressure on the Behavior of Material

When the temperature remains constant, the pressure also affects the behavior of a material. The volume of a gas changes as the pressure changes even if temperature remains the same. When temperature is close to the melting point of a substance, a liquid may solidify or a solid may melt as the pressure changes. A diagram showing the temperature and pressure combined effect on a system is called a phase diagram.

One-component phase diagrams for water and carbon dioxide are given here.

At pressure below 5.1 atm, solid and gas carbon dioxide coexist, but the vapor pressure depends on the temperature. The variation of vapor pressure is represented by a line, which separates the Dry ice phase from the CO2 Gas phase. The vapor pressure of dry ice at 194.6 K (-78.5°C) the pressure is 1 atm, and at 216.7 K (-56.4°C) the pressure is 5.11 atm. The line separates the conditions for the formation of solid and vapor. A similar curve borders between liquid and gas CO2, whereas a line separates dry ice from the liquid phase. At 216.7 K, vapor pressures of solid and liquid CO2 are the same, 5.11 atm. At, 5.11 atm and 216.7 K, all three phases coexist, and the condition is called the triple point.

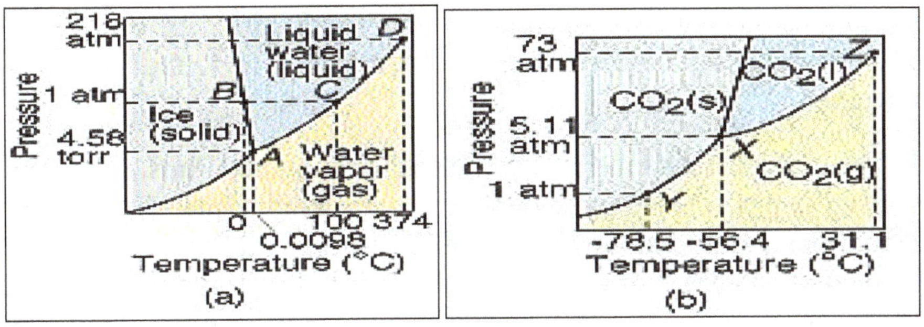

Thermal Expansion Coefficient

A substance expands on heating. For a rod, the lengthening of a unit length per degree Kelvin is the linear thermal expansion coefficient. This factor affects the substance performance in machines or structural assemblies. Thermal expansion causes tight fitted parts to break and moving part to jam, in any machine. The problem is serious if different material is used. When a large body of glass is subject to local heating or cooling, it breaks up due to expansion or shrinkage. Thermal expansion also causes distortion, and some thermometers are made of two strips of different metals. Thermal properties must be considered in any engineering constructions such as railroad, bridges, pipelines, and buildings, especially in areas where temperatures go to extreme values.

Linear thermal expansion coefficient per K at room temperature of some substances	
Aluminum	24
Copper	17
Diamond	1
Glass	11
Pyrex glass	3
Rubber, hard	80

Electric resistance of some Elements	
Metal	Resistance /ohm m
Al	$2.4x10^{-8}$
Au	$2.05x10^{-8}$
W	$5.65x10^{-6}$
Ge	0.46
Si	0.03
S (yellow)	$2x10^{15}$
P (white)	109 Note the range of 10^{15}

Heat and Electric Conductance

Transmissions of energy and electric charge across a body of material give rise to heat and electric conductance respectively. The rate of flow across a unit-area section when the temperature or electric potential difference applied to the wire of unit length is called the thermal or electric conductance coefficient. Metals are usually good conductors of both, and their conductance coefficients are high. Insulation material for heat and electricity should have low conductance, whearas metals have high conductance.

The reciprocal of electric conductance is called electric resistance; thus, the higher the conductance, the lower the resistance. Electric resistance for some familiar materials are given in the table here. Note the large range of 10^{15} among these substances. Aluminium and copper are very good conductors, and their resistances are very low, in the order of 10^{-8}, almost 100 times smaller than that of tungsten, W. Germanium, Ge, and silicon, Si, are typical semiconductors, whereas sulfur and phosphorous are insulation material.

Magnetic Properties

A magnetic field strength is measured in Tesla (T) and gauss (G, 1 T = 10,000 G). The Earth magnetic field is 0.5 G. When a material is placed into a magnetic field H, a magnetic field of different intensity B is produced inside the material. The ratio B/H is called the magnetic susceptibility. The higher the magnetic susceptibility, the easier the material is magnetized. Most substances are diamagnetic. The magnetic fields (B) within the bodies of these substances when they are placed in a magnetic field (H) are less than that of an empty space (vacuum).; thus their magnetic susceptibilities (B/H ratio) are less than 1. When a body of paramagnetic substance is placed in a magnetic field, the intensity of the field within the body is slightly larger than that of the applied field. The magnetic susceptibilities of paramagnetic substances are slightly greater than 1.

Iron, cobalt and nickel are some ferromagnetic substances, there are some other alloys and oxides that behave this way. They possess a spontaneous magnetic moment. A magnetic field is present in these materials even in the absence of an external magnetic field. However, ferromagnetism is temperature dependent, and above the so called Curie temperatures of the substances, magnetism vanishes. The Curie temperature or Curie point of a substance is unique. The Curie points for Fe, Co, and Ni are 1043, 1400, and 630 K respectively.

Ferromagnetism are due to the presence of magnetic domains in the substance, and when these domains line up parallel to each other, they give a net magnetic field. If the domains line up antiparallel to each other at the Curie point, the substance is said to be antiferromagnetic. The magnetic susceptibility reaches a maximum at Curie temperature for antferromagnetic material. For example, F_eO, MnO, C_oO, N_iO, F_eF_2, F_eCl_2, a-Mn, $C_{r2}O_3$ etc. are some of the antiferromagnetic substances.

Ferromagnetic substances play important roles in recording tapes and disks for audio, video, and computer signals. Furthermore, ferromagnetic materials are used in permanent magnets, which are used in motors, antenna, and speakers. Recent development in strong magnets enables communication equipment and computers to be miniaturized.

Density

The mass per unit volume (cm^3 = mL, m^3 etc.) of is called density, an intensive property. Often, specific gravity is given. Specific gravity is the ratio of density of a substance compared to that of water. As a ratio, it has no units. Since density of water is 1.00 g/mL, specific gravity is the density in g/mL. Other units to use are kg/L or 10^3 kg m^{-3}. Specific gravity for a few common substances are given here: Au, 19.3; mercury, 13.6; alcohol, 0.7893; benzene, 0.8786. Do you know which element has the highest density?

Dielectric Constant

The dielectric constant e of a medium is its ability to reduce the force F of attraction of charged (q_1 and q_2) particles separated at distance r, compared to vacuum. It is usually defined by the equation, $F = q_1 q_2 / (e\ r)$. A substance with large dielectric constant placed between two plates to which an electric voltage has been applied will result in a weak electric field within it. Water, due to its polar nature, has a rather large dielectric constant, 80.4. At the atomic scale, water molecules weaken the attraction between Na+ and Cl- ions, resulting in dissolving it. Dielectric constants for some familiar substances are: H_2O, 80.4; methanol, 33.6; benzene, 2.3; H_2 at 20 K, 1.23.

Heat Capacity

The amount of energy required to raise the temperature of a substance by 1 K is the heat capacity. If the substance has a unit mass, the amount is referred to as specific heat capacity, or specific heat. For example, it takes 1 cal (4.184 J) to raise the temperature of 1 g water by 1 K. Thus, the specific heat for water is 1 cal g^{-1} K^{-1} (75 J mol^{-1} K^{-1}). Specific heat of water is large compared to most other substances, for example: Cu, 24.4 J mol^{-1} K^{-1}. This large heat capacity of water affect the weather, making temperatures in areas close to large bodies of water more steadier than large dry land.

Refractive Index

Refractive index of some common substances	
water	1.3
benzene	1.5011
ethanol	1.359
quartz	1.5*
NaCl solid	1.5
*wavelength dependent	

The ratio of light speed in vacuum to its speed in the medium is refractive index. Light travel slower in any medium than in vacuum. Thus, refractive index is always greater than unity (1), and light beam usually bents when entering from air to another medium. This value depends on the wavelength of the light used, and the property is important for material used in optics instrument such as eye glasses. The higher the refractive index, the thinner the glasses to achieve the same power. The difference in refractive indexes between two liquid gives rise to the visible boundary between layers.

Difference in refractive indexes of lights of different wavelengths can be separated using a prism. Refractive indexes for some familiar substances are given in a box. It should also be kept in mind that index of refraction changes with dissolved substance and concentration.

Solubility

The amount of substance dissolved in 100 mL of solvent is called solubility. However, units for solubility might be specified in some other fashion. Solubility depends on temperature, and the variation can be used to separate components in a solid mixture. Sodium acetate trihydrate, $CH_3COONa{\cdot}3H_2O$, when heated will melt in the sense that it dissolves in its water of crystallization. This liquid remains liquid till about -15 C (258 K), and when crystallization does take place after triggered by cold hand, heat is released providing a source of heat. This property provides a winter hand warmer pack for skiers or winter travelers.

Washing and cleaning also involve solubility, and the formulation of an effective cleaning agent requires the knowledge of many substances. Substances can be classified according to polarity. Water, ammonia (NH_3), and methanol (CH_3OH) are polar, because their molecules have negative and positive sites, whereas methane (CH_4), gasoline, and motor oil are non-polar. Regarding solubility, a rule of thumb reads like dissolves like, which means that polar solvents dissolve polar substances and non-polar solvents dissolve non-polar substances. An organic compound with a polar group attached to non-polar chain will bring water molecules to a non-polar surface, and hence it can be used as a detergent or wetting agent. This rule of thumb has potential for both domestic and industrial applications.

Optical Activity

The ability of certain substances to rotate the plane of polarized light as it passes through a crystal, liquid, or solution is generally referred to as the optical activity. Substances possessing this activity usually lack a center of symmetry, and they have two isomers as mirror images of each other. The two isomers, called dextrorotatory (d, right hand) or laevorotory (l, left hand) isomers, rotate the polarized light in opposite directions. Thus, equi-molar or racemic mixture of the two appears optically inactive. For example, sugar, tartaric acid, and aminoacids are optically active compounds.

Viscosity and Surface Tension

Viscosity and surface tension are properties of liquid state. The former is a measure of its resistance to flow. Molasses, glycerin, oil, softened glass have high viscosity, and water, gasoline, ethanol have low viscosity. The SI units for viscosity is N.s /m2, but the unit poise (P, a cgs unit) have been used for a long time, and is more common, and 1 P = 0.1 N¸s m^{-2}. Viscosity usually decreases with increase in temperature, and softened glass has a viscosity more than 1014 N¸s m^{-2}.

Surface tension results from intermolecular attraction, the higher of which, the higher surface tension. Energy required to stretch the surface by some defined unit is called surface tension, and whose unit is N.m/m^2 (= N/m). Like viscosity, surface tension decrease with increase temperature. Surface tension causes the dew and raindrops to be round, and the idea to manufacture perfect spheres in zero gravity zone is making use of surface tension. Soap reduces surface tension of water and soapy water forms bubbles when air is blown into it.

Electrical Properties

It is well known that one of the subatomic particles of an atom is the electron. The electrons carry a negative electrostatic charge and under certain conditions can move from atom to atom. The direction of movement between atoms is random unless a force causes the electrons to move in one direction. This directional movement of electrons due to an electromotive force is what is known as electricity.

Electrical Conductivity

Electrical conductivity is a measure of how well a material accommodates the movement of an electric charge. It is the ratio of the current density to the electric field strength. Its SI derived unit is the Siemens per meter, but conductivity values are often reported as percent IACS. IACS is an acronym for International Annealed Copper Standard, which was established by the 1913 International Electrochemical Commission. The conductivity of the annealed copper (5.8001 x 107S/m) is defined to be 100% IACS at 20°C . All other conductivity values are related back to this conductivity of annealed copper. Therefore, iron with a conductivity value of 1.04 x 107 S/m, has a conductivity of approximately 18% of that of annealed copper and this is reported as 18% IACS. An interesting side note is that commercially pure copper products now often have IACS conductivity values greater than 100% IACS because processing techniques have improved since the adoption of the standard in 1913 and more impurities can now be removed from the metal.

Conductivity values in Siemens/meter can be converted to % IACS by multiplying the conductivity value by 1.7241×10^{-6}. When conductivity values are reported in microSiemens/centimeter, the conductivity value is multiplied by 172.41 to convert to the % IACS value.

Electrical conductivity is a very useful property since values are affected by such things as a substances chemical composition and the stress state of crystalline structures. Therefore, electrical conductivity information can be used for measuring the purity of water, sorting materials, checking for proper heat treatment of metals, and inspecting for heat damage in some materials.

Electrical Resistivity

Electrical resistivity is the reciprocal of conductivity. It is the is the opposition of a body or substance to the flow of electrical current through it, resulting in a change of electrical energy into heat, light, or other forms of energy. The amount of resistance depends on the type of material. Materials with low resistivity are good conductors of electricity and materials with high resistivity are good insulators.

The SI unit for electrical resistivity is the ohm meter. Resistivity values are more commonly reported in micro ohm centimeters units. As mentioned above resistivity values are simply the reciprocal of conductivity so conversion between the two is straightforward. For example, a material with two micro ohm centimeter of resistivity will have ½ microSiemens/centimeter of conductivity. Resistivity values in microhm centimeters units can be converted to % IACS conductivity values with the following formula:

$$172.41 / \text{resistivity} = \%\ \text{IACS}$$

Temperature Coefficient of Resistivity

Electrical conductivity values (and resistivity values) are typically reported at 20° C. This is done because the conductivity and resistivity of material is temperature dependant. The conductivity of most materials decreases as temperature increases. Alternately, the resistivity of most material increases with increasing temperature. The amount of change is material dependant but has been established for many elements and engineering materials.

The reason that resistivity increases with increasing temperature is that the number of imperfection in the atomic lattice structure increases with temperature and this hampers electron movement. These imperfections include dislocations, vacancies, interstitial defects and impurity atoms. Additionally, above absolute zero, even the lattice atoms participate in the interference of directional electron movement as they are not always found at their ideal lattice sites. Thermal energy causes the atoms to vibrate about their equilibrium positions. At any moment in time many individual lattice atoms will be away from their perfect lattice sites and this interferes with electron movement.

When the temperature coefficient is known, an adjusted resistivity value can be computed using the following formula:

$$R_1 = R_2 * [1 + a * (T_1 - T_2)]$$

Where: R_1 = resistivity value adjusted to T_1

R_2 = resistivity value known or measured at temperature T2

a = Temperature Coefficient

T_1 = Temperature at which resistivity value needs to be known

T_2 = Temperature at which known or measured value was obtained

For example, suppose that resistivity measurements were being made on a hot piece of aluminum. Normally when measuring resistivity or conductivity, the instrument is calibrated using standards that are at the same temperature as the material being measured, and then no correction for temperature will be required. However, if the calibration standard and the test material are at different temperatures, a correction to the measured value must be made. Presume that the instrument was calibrated at 20°C (68°F) but the measurement was made at 25°C (77°F) and the resistivity value obtained was 2.706×10^{-8} ohm meters. Using the above equation and the following temperature coefficient value, the resistivity value corrected for temperature can be calculated.

$$R_1 = R_2 * [1 + a * (T_1 - T_2)]$$

Where: R_1 = ?

$R_2 = 2.706 \times 10^{-8}$ ohm meters (measured resistivity at 25°C)

a = 0.0043/ °C

T1 = 20°C

T2 = 25°C

$$R1 = 2.706 \times 10^{-8} \text{ohm meters} * [1 + 0.0043/\text{oC} * (20\ ^\circ\text{C} - 25\ ^\circ\text{C})]$$

$$R1 = 2.648 \times 10^{-8} \text{ohm meters}$$

Note that the resistivity value was adjusted downward since this example involved calculating the resistivity for a lower temperature.

Since conductivity is simply the inverse of resistivity, the temperature coefficient is the same for conductivity and the equation requires only slight modification. The equation becomes:

$$s_1 = s_2 / [1 + a * (T_1 - T_2)]$$

Where: s_1 = conductivity value adjusted to T_1

s_2 = conductivity value known or measured at temperature T_2

a = Temperature Coefficient

T_1 = Temperature at which conductivity value needs to be known

T_2 = Temperature at which known or measured value was obtained

In this example let's consider the same aluminum alloy with a temperature coefficient of 0.0043 per degree centigrade and a conductivity of 63.6% IACS at 25°C. What will the conductivity be when adjusted to 20°C?

s_1= 63.6% IACS / [1 + 0.0043 * (20°C – 25°C)]

s_1= 65.0% IASC

The temperature coefficient for a few metallic elements is shown below.

Material	Temperature Coefficient (/°C)
Nickel	0.0059
Iron	0.0060
Molybdenum	0.0046
Tungsten	0.0044
Aluminum	0.0043
Copper	0.0040
Silver	0.0038
Platinum	0.0038
Gold	0.0037
Zinc	0.0038

Energy Band Structures in Solids

When atoms come together to form a solid, their valence electrons interact due to Coulomb forces, and they also feel the electric field produced by their own nucleus and that of the other atoms. In addition, two specific quantum mechanical effects happen. First, by Heisenberg's uncertainty principle, constraining the electrons to a small volume raises their energy, this is called promotion. The second effect, due to the Pauli exclusion principle, limits the number of electrons that can have the same property (which include the energy). As a result of all these effects, the valence electrons of atoms form wide valence bands when they form a solid. The bands are separated by gaps, where electrons cannot exist. The precise location of the bands and band gaps depends on the type of atom (e.g., Si vs. Al), the distance between atoms in the solid, and the atomic arrangement (e.g., carbon vs. diamond).

In semiconductors and insulators, the valence band is filled, and no more electrons can be added, following Pauli's principle. Electrical conduction requires that electrons be able to gain energy in an electric field; this is not possible in these materials because that would imply that the electrons are promoted into the forbidden band gap.

In metals, the electrons occupy states up to the Fermi level. Conduction occurs by promoting electrons into the conduction band, that starts at the Fermi level, separated by the valence band by an infinitesimal amount.

Conduction in Terms of Band and Atomic Bonding Models

Conduction in metals is by electrons in the conduction band. Conduction in insulators is by electrons in the conduction band and by holes in the valence band. Holes are vacant states in the valence band that are created when an electron is removed.

In metals there are empty states just above the Fermi levels, where electrons can be promoted. The promotion energy is negligibly small so that at any temperature electrons can be found in the conduction band. The number of electrons participating in electrical conduction is extremely small.

In insulators, there is an energy gap between the valence and conduction bands, so energy is needed to promote an electron to the conduction band. This energy may come from heat, or from energetic radiation, like light of sufficiently small wavelength.

A working definition for the difference between semiconductors and insulators is that in semiconductors, electrons can reach the conduction band at ordinary temperatures, where in insulators they cannot. The probability that an electron reaches the conduction band is about $\exp(-E_g/2kT)$ where E_g is the band gap and kT has the usual meaning. If this probability is, say, $< 10^{-24}$ one would not find a single electron in the conduction band in a solid of 1 cubic centimeter. This requires $E_g/2kT > 55$. At room temperature, $2kT = 0.05$ eV; thus $E_g > 2.8$ eV can be used as the condition for an insulator.

Besides having relatively small E_g, semiconductors have covalent bond, whereas insulators usually are partially ionic bonded.

Electron Mobility

Electrons are accelerated in an electric field E, in the opposite direction to the field because of their negative charge. The force acting on the electron is -eE, where e is the electric charge. This force produces a constant acceleration so that, in the absence of obstacles (in vacuum, like inside a TV tube) the electron speeds up continuously in an electric field. In a solid, the situation is different. The electrons scatter by collisions with atoms and vacancies that change drastically their direction of motion. Thus electrons move randomly but with a net drift in the direction opposite to the electric field. The drift velocity is constant, equal to the electric field times a constant called the mobility μ,

$$v_d = -\mu_e E$$

which means that there is a friction force proportional to velocity. This friction translates into energy that goes into the lattice as heat. This is the way that electric heaters work.

The electrical conductivity is:

$$\sigma = n|e|\mu_e$$

where n is the concentration of electrons (n is used to indicate that the carriers of electricity are negative particles).

Electrical Resistivity of Metals

The resistivity then depends on collisions. Quantum mechanics tells us that electrons behave like waves. One of the effects of this is that electrons do not scatter from a perfect lattice. They scatter by defects, which can be:

- Atoms displaced by lattice vibrations.
- Vacancies and interstitials.
- Dislocations, grain boundaries.
- Impurities.

One can express the total resistivity ρ_{tot} by the Matthiessen rule, as a sum of resistivities due to thermal vibrations, impurities and dislocations. Illustrates how the resistivity increases with temperature, with deformation, and with alloying.

Electrical Characteristics of Commercial Alloys

The best material for electrical conduction (lower resistivity) is silver. Since it is very expensive, copper is preferred, at an only modest increase in ρ. To achieve low r it is necessary to remove gases occluded in the metal during fabrication. Copper is soft so, for applications where mechanical strength is important, the alloy CuBe is used, which has a nearly as good ρ. When weight is important one uses Al, which is half as good as Cu. Al is also more resistant to corrosion.

When high resistivity materials are needed, like in electrical heaters, especially those that operate at high temperature, nichrome (N_iC_r) or graphite are used.

Intrinsic Semiconduction

Semiconductors can be intrinsic or extrinsic. Intrinsic means that electrical conductivity does not depend on impurities, thus intrinsic means pure. In extrinsic semiconductors the conductivity depends on the concentration of impurities.

Conduction is by electrons and holes. In an electric field, electrons and holes move in opposite direction because they have opposite charges. The conductivity of an intrinsic semiconductor is:

$$\sigma = n|e|\mu_e + \rho|e|\mu_h$$

where ρ is the hole concentration and μ_h the hole mobility. One finds that electrons move much faster than holes:

$$\mu_e > \mu_h$$

In an intrinsic semiconductor, a hole is produced by the promotion of each electron to the conduction band. Thus:

$$n = p$$

Thus, $\sigma = 2n|e|\ (\mu_e + \mu_h)$ (only for intrinsic semiconductors).

Extrinsic Semiconduction

Unlike intrinsic semiconductors, an extrinsic semiconductor may have different concentrations of holes and electrons. It is called p-type if p>n and n-type if n>p. They are made by doping, the addition of a very small concentration of impurity atoms. Two common methods of doping are diffusion and ion implantation.

Excess electron carriers are produced by substitutional impurities that have more valence electron per atom than the semiconductor matrix. For instance phosphorous, with 5 valence electrons, is an electron donor in Si since only 4 electrons are used to bond to the Si lattice when it substitutes for a Si atom. Thus, elements in columns V and VI of the periodic table are donors for semiconductors in the IV column, Si and Ge. The energy level of the donor state is close to the conduction band, so that the electron is promoted (ionized) easily at room temperature, leaving a hole (the ionized donor) behind. Since this hole is unlike a hole in the matrix, it does not move easily by capturing electrons from adjacent atoms. This means that the conduction occurs mainly by the donated electrons (thus n-type).

Excess holes are produced by substitutional impurities that have fewer valence electrons per atom than the matrix. This is the case of elements of group II and III in column IV semiconductors, like B in Si. The bond with the neighbors is incomplete and so they can capture or accept electrons from adjacent silicon atoms. They are called acceptors. The energy level of the acceptor is close to the valence band, so that an electron may easily hop from the valence band to complete the bond leaving a hole behind. This means that conduction occurs mainly by the holes (thus p-type).

Dielectric Behavior

A dielectric is an electrical insulator that can be made to exhibit an electric dipole structure (displace the negative and positive charge so that their center of gravity is different).

Capacitance

When two parallel plates of area A, separated by a small distance l, are charged by +Q, –Q, an electric field develops between the plates,

$$E = D/\varepsilon\varepsilon_o$$

where $D = Q/A$. ε_o is called the vacuum permittivity and e the relative permittivity, or dielectric constant (e = 1 for vacuum). In terms of the voltage between the plates, $V = E\,l$,

$$V = Dl/\varepsilon\varepsilon_o = Q\,l/A\varepsilon\varepsilon_o = Q / C$$

The constant $C= A\varepsilon\varepsilon_o/l$ is called the capacitance of the plates.

The dipole moment of a pair of positive and negative charges (+q and –q) separated at a distance d is p = qd. If an electric field is applied, the dipole tends to align so that the positive charge points in the field direction. Dipoles between the plates of a capacitor will produce an electric field that opposes the applied field. For a given applied voltage V, there will be an increase in the charge in the plates by an amount Q' so that the total charge becomes $Q = Q' + Q_o$, where Q_o is the charge of a vacuum capacitor with the same V. With Q' = PA, the charge density becomes $D = D_0 E + P$, where the polarization $P = \varepsilon_0 (\varepsilon - 1) E$.

Types of Polarization

Three types of polarization can be caused by an electric field:

- Electronic polarization: the electrons in atoms are displaced relative to the nucleus.
- Ionic polarization: cations and anions in an ionic crystal are displaced with respect to each other.
- Orientation polarization: permanent dipoles (like H2O) are aligned.

Frequency Dependence of the Dielectric Constant

Electrons have much smaller mass than ions, so they respond more rapidly to a changing electric field. For electric field that oscillates at very high frequencies (such as light) only electronic polarization can occur. At smaller frequencies, the relative displacement of positive and negative ions can occur. Orientation of permanent dipoles, which require the rotation of a molecule can occur only if the oscillation is relatively slow (MHz range or slower). The time needed by the specific polarization to occur is called the relaxation time.

Dielectric Strength

Very high electric fields (>108 V/m) can free electrons from atoms, and accelerate them to such high energies that they can, in turn, free other electrons, in an avalanche process (or electrical discharge). This is called dielectric breakdown, and the field necessary to start the is called the dielectric strength or breakdown strength.

Dielectric Materials

Capacitors require dielectrics of high e that can function at high frequencies (small relaxation times). Many of the ceramics have these properties, like mica, glass, and porcelain). Polymers usually have lower e.

Ferroelectricity

Ferroelectric materials are ceramics that exhibit permanent polarization in the absence of an electric field. This is due to the asymmetric location of positive and negative charges within the unit cell. Two possible arrangements of this asymmetry results in two distinct polarizations, which can be used to code "0" and "1" in ferroelectric memories. A typical ferroelectric is barium titanate, $BaTiO_3$, where the Ti^{4+} is in the center of the unit cell and four O^{2-} in the central plane can be displaced to one side or the other of this central ion.

Piezoelectricity

In a piezolectric material, like quartz, an applied mechanical stress causes electric polarization by the relative displacement of anions and cations.

Biological Properties

A living organism has a material structure to provide an environment for complicated chemistry of living. Chemical and physical reactions provide energy to maintain living functions and to renew structural material. Thus, consideration of biological properties is a natural extension of physical and chemical properties.

To a large extend, biological functions of any materials are related to their chemical and physical properties. However, reactions in biological systems are catalyzed by enzymes. Furthermore, products of one reaction may be reactants for another in a complicate scheme of reactions to maintain live. Malfunction of a reaction causes trouble, leading to disease or death. Thus, biological properties deserve special consideration.

Absorption and Transport

Generally speaking, food, medicine, and toxin can be given to a person by means of feeding, absorption, inhalation and injection. Food must be absorbed in the digestive track and transported from to the targeted organs, tissues and cells to provide energy for life. Injected substances may already be in organs or tissues, but they may not be in the targeted cells. In any cases, materials in question must pass through membranes, either by enzyme aided (active) transport or by diffusion (passive transport). Transportation of a substance in a biological system is a very complicated process.

Absorption and transport of a substance often involve its solubility in the medium. Substances prefer to dissolve in water type fluid are said to be hydrophilic, whereas those prefer to dissolve in oily fluids are lipophilic. Absorption and transport of substances depend on their solubility in water and lipid media. The ultimate effect on cells, tissues, and organs must take place at the molecular level. However, effects at the molecular level are often not observable, and the symptoms of these effects may appear to be unrelated to the material in question.

Solid substances such as fiber, gold and charcoal, not absorbed and used in any biological function pass out as feces. Absorbed soluble substances, but not utilized by animals are excreted with urine.

Material for Biological Structure

Bullets and shell fragments anchored in bones of war veterans were not removed, because doctors considered them biologically inert. Stainless steel parts replace joints and bones today, because they are inert.

Both inorganic and organic materials are involved in structures of living organism. Lignin, cellulose, muscles, skins, and cell walls are mostly organic, whereas bones, teeth, and shells involve mostly inorganic substances. These substances may serve only as structural materials in biological systems, and if so they can be replaced by biological inert substances.

However some subsystems of structural organs are responsible for vital biological functions. For example, bone marrow is responsible for blood regeneration. Thus, replacement of biological structure material involves many disciplines.

Biological Materials and Biomaterial

Plasma, membrane, tissue, protein, lipid, enzyme, the digestive system, and the central nervous systems are some examples of biological materials, for which properties for consideration include growth and decay, turn over time, biological half-life, retention time, composition and its change, and active ingredient. These are manifestation of physical and chemical properties of biological materials. However, biological properties allow us to identify and solve the biological problems. Biological materials had been studied by biologists, chemists, and engineers from the macroscopic, molecular, and functional viewpoints.

In contrast, replacement or implant materials that imitate living tissues or organs are called biomaterial, which can be divided into two categories: soft tissue and hard tissue replacement biomaterial. The former includes sutures, surgical tapes, adhesives, skin implants etc. The latter include metals (steel, aluminum, titanium, cobalt-based alloys, and titanium-based alloys); ceramics (made up of Al_2O_3, TiO_2, SiO_2, Fe_2O_3 etc.); carbon (graphite and glassy carbon); and polymers.

The chemistry of living is complex, and properties of biological materials towards biomaterials are of great interest. The general reaction of biological materials towards foreign biomaterials is expel (or rejection). Living tissues form a thin layer around the inert biomaterial, but materials that irritate the tissues causes inflammation. Most pure metals evoke severe tissue reaction due to their redox reactions. However, aluminum and titanium are metals of choice, because the formation of a thin oxide layer on their surface made them inert. Similarly, ceramics are compatible to body fluid because they are made of the metal oxides. The nature of the surface also affects the biological properties, rough ones enable tight attachment of tissues.

Biological Activity

Biological activities of materials can be divided according to biological functions. Substances that provide nutrition, energy, and structural need are called food, whereas those that disrupt the normal functions are called toxins. Substances used to correct the abnormal biological functions are called medicines.

In recent years, a lot of research had gone into finding quantitative structure-activity relationships (QSAR) of various substances aimed at improving drug design. As they provide an indication of some biological activity, we list some categories here:

- Anti-infective agents,
- Antibiotics,
- Anti-tumor agents,
- Cardiovascular agents,
- Anti-allergic,
- Anti-ulcer,
- Anti-inflammatory, and
- Anti-arthritics.

Agents affecting the central nervous system includes analgesics, anesthetics, antidepressants, convulsants, anti-convulsants, neuroleptics, and psychotomimetics. There are also steroids and hormones which interacts with genes in the nuclei of cells causing complicated developments.

Mechanical Properties

The mechanical properties of a material are those which effect the mechanical strength and ability of a material to be molded in suitable shape. Some of the typical mechanical properties of a material are listed below-

- Strength
- Toughness
- Hardness
- Hardenability
- Brittleness
- Malleability
- Ductility
- Creep and Slip
- Resilience
- Fatigue

Strength

It is the property of a material which opposes the deformation or breakdown of material in presence of external forces or load. Materials which we finalize for our engineering products, must have suitable mechanical strength to be capable to work under different mechanical forces or loads.

Toughness

It is the ability of a material to absorb the energy and gets plastically deformed without fracturing. Its numerical value is determined by the amount of energy per unit volume. Its unit is Joule/ m3. Value of toughness of a material can be determined by stress-strain characteristics of a material. For good toughness, materials should have good strength as well as ductility.

For example: brittle materials, having good strength but limited ductility are not tough enough. Conversely, materials having good ductility but low strength are also not tough enough. Therefore, to be tough, a material should be capable to withstand both high stress and strain.

Hardness

It is the ability of a material to resist to permanent shape change due to external stress. There are various measure of hardness – Scratch Hardness, Indentation Hardness and Rebound Hardness.

- Scratch Hardness:

  Scratch Hardness is the ability of materials to the oppose the scratches to outer surface layer due to external force.

- Indentation Hardness:

  It is the ability of materials to oppose the dent due to punch of external hard and sharp objects.

- Rebound Hardness:

  Rebound hardness is also called as dynamic hardness. It is determined by the height of "bounce" of a diamond tipped hammer dropped from a fixed height on the material.

Hardenability

It is the ability of a material to attain the hardness by heat treatment processing. It is determined by the depth up to which the material becomes hard. The SI unit of hardenability is meter (similar to length). Hardenability of material is inversely proportional to the weld-ability of material.

Brittleness

Brittleness of a material indicates that how easily it gets fractured when it is subjected to a force or load. When a brittle material is subjected to a stress it observes very less energy and gets fractures without significant strain. Brittleness is converse to ductility of material. Brittleness of material is temperature dependent. Some metals which are ductile at normal temperature become brittle at low temperature.

Malleability

Malleability is a property of solid materials which indicates that how easily a material gets deformed under compressive stress. Malleability is often categorized by the ability of material to be formed in the form of a thin sheet by hammering or rolling. This mechanical property is an aspect of plasticity of material. Malleability of material is temperature dependent. With rise in temperature, the malleability of material increases.

Ductility

Ductility is a property of a solid material which indicates that how easily a material gets deformed under tensile stress. Ductility is often categorized by the ability of material to get stretched into a wire by pulling or drawing. This mechanical property is also an aspect of plasticity of material and is temperature dependent. With rise in temperature, the ductility of material increases.

Creep and Slip

Creep is the property of a material which indicates the tendency of material to move slowly and deform permanently under the influence of external mechanical stress. It results due to long time exposure to large external mechanical stress with in limit of yielding. Creep is more severe in material that are subjected to heat for long time. Slip in material is a plane with high density of atoms.

Resilience

Resilience is the ability of material to absorb the energy when it is deformed elastically by applying stress and release the energy when stress is removed. Proof resilience is defined as the maximum energy that can be absorbed without permanent deformation. The modulus of resilience is defined as the maximum energy that can be absorbed per unit volume without permanent deformation. It can be determined by integrating the stress-strain cure from zero to elastic limit. Its unit is joule/m^3.

Fatigue

Fatigue is the weakening of material caused by the repeated loading of the material. When a material is subjected to cyclic loading, and loading greater than certain threshold value but much below the strength of material (ultimate tensile strength limit or yield stress limit), microscopic cracks begin to form at grain boundaries and interfaces. Eventually the crack reaches to a critical size. This crack propagates suddenly and the structure gets fractured. The shape of structure affects the fatigue very much. Square holes and sharp corners lead to elevated stresses where the fatigue crack initiates.

Thermal Properties

A thermal property iis any characteristic of a material defining the substance and related to temperature; e.g. thermal conductivity is said to be a thermal property, but electrical conductivity is not. However, all properties, thermal and non-thermal, are temperature dependent, and in this sense included under thermal properties.

The effect of temperature on thermal properties may be large (what may be used to build good thermometers). Standard values are usually given at 20 °C (comfort lab conditions), but other reference conditions are also traditionally used: 0 °C because its ease of reproducing, 15 °C because it is the average temperature in the Earth surface, 20 °C (human comfort), or 25 °C because it is easier to maintain a bath temperature a little over the oscillating ambient temperature, than below. Fortunately, the influence of all those temperature-standards is minor on property values, but care should be paid to make it explicit.

The effect of pressure on thermal properties is very low on condense substances. The standard value for pressure is 100 kPa, although 101.325 kPa, the average pressure in the Earth surface, is sometimes used. The effect of uncertainty in composition of the substance is usually small (e.g. properties of tap water, and even of sea water, may be taken as those of pure water, in many instances), except on some sensitive properties, like for the thermo-optical properties of substances, that are heavily dependent on contamination, or the thermal conductivity of metals, that may vary a lot with small alloys, etc.

Specific Heat Capacity

Specific heat capacity of inorganic building materials (concrete, break, natural stone) is varied within 0.75-0.92 kJ/(kg °C), for wood it is 0.7 kJ/(kg °C). Since water has very high specific heat capacity of 4 kJ/(kg °C), the specific heat capacity of materials increases with the increase of their humidity. Specific heat capacity of composite materials can vary significantly if temperature variations are accompanied by phase changes. Typical example is related to sea ice consisting fresh ice and brine cells. The heating of sea ice is accompanied not only by the increasing of sea ice temperature but also by the melting of fresh ice around brine cells. This process becomes more important with increasing sea ice temperature.

	Solids	Liquids
Specific heat capacity, kJ/(kg°C)		Ammonia (4.19); Water (4)
	Plastic (1.76; Cork, Rubber (1.68); Wool (1.63); Cellulose (1.55); Coal, Naphthalene (1.3); Concrete (1.13)	Aquafortis (2.77); Hexane (2.51); Phenol (2.35); Kerosene (2.1); Azote liquid (2.01)
		Benzene (1.84); Turpentine (1.7); Luboil, Liquid oxygen (1.68); Nitrobenzene (1.38)
	Aluminum, Clay, Brick (0.92); Coke (0.84); Sand (0.8); Glass (0.84-0.42); Slag (0.75); Wood (0.7); Iron, Steel (0.5); Copper (0.385); Zinc (0.38); Lead (0.13	

Table: Mean specific heat capacities of solid and liquid materials in temperature range 0-100° C.

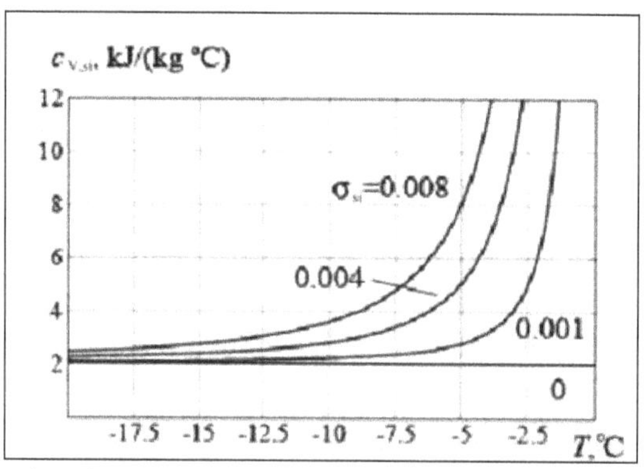

Figure: Specific heat capacity of sea ice versus its temperature for different values of sea ice salinity.

Thermal Conductivity

Thermal conductivity of materials with simple chemical composition is greater than thermal conductivity of materials of complicated chemical composition. Thermal conductivity of materials with crystal structure is higher than thermal conductivity of materials with mixed or amorphous structure. For example, mean thermal conductivities of single crystal of quartz is 7-8 W/(m °C), for sand-rock with impurities it is 2.1-2.9 W/(m °C), and for normal glass with amorphous structure it is 0.76 W/(m °C). Porous materials conduct Porous material has smaller thermal conductivity than continuous materials when the pores are filled by air. Thermal conductivity of materials with small closed pores is smaller than thermal conductivity of the same material with bigger pores under the same overall porosity. It is because heat transfer due to convection is reduced in the material with smaller pores.

Material	Density, kg/m³	Thermal conductivity, W/(m°C)
Foam Plastic	30	0.047
Cork fines	110	0.047
Glass-woo	200	0.35-0.047
Cinder-woo	250	0.076
Felt-wool	300	0.47
Wood cross fibers	600	0.14-0.174
Wood along fibers	600	0.384
Asbestos	600	0.151
Insulation brick	600	0.116-0.209
Textolite	1380	0.244
Dry sand	1500	0.349-0.814
Brick lining	1700	0.698-0.814
Fire brick	1840	1.05
Concrete	2300	1.28

Aluminum	2700	203.5
Cast Iron	7500	46.5-93
Steel	7850	36.5
Stainless steel	7900	17.5
Bronze	8000	64
*Latten	8500	93
Copper	8800	384
Lead	11400	34.9

*Bronze-like yellow alloy used to make church utensils in the middle ages by beating it into thin sheets by virtue of its malleability and ductility.

Melting Point

Crystal materials have certain melting points above which their crystal structure is destroyed. Below their melting point crystal materials are solid and above it they become liquids. The softening of amorphous materials occurs gradually with increasing temperature, evolving into viscous fluids with decreasing viscosity under increasing temperature.

Table: Melting points if some materials.

Materi	Melting point, o C	Material	Melting Point .C
Water	0	Zinc	419
Wolfram	3370	Lead	327
Gold	1063	Tin	232
Iron	1535	Mercury	39
Copper	1083		

Physical properties of many materials depend on the proximity of their temperature to the melting point. This property is characterized by homologous temperature calculated as a ratio of the actual temperature to the melting point. Homologous temperatures of some materials are shown in figure b) in natural range of actual temperature from -100 °C to 100 °C.

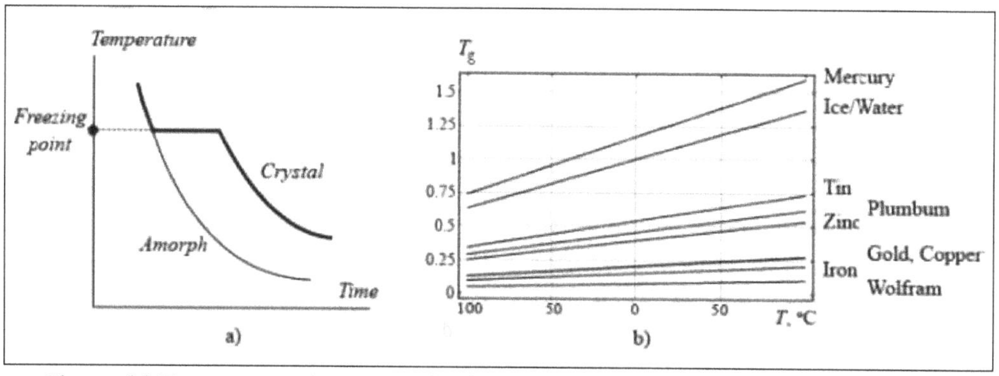

Figure: (a).Temperature-time curves for the cooling of amorphous and crystal materials, (b). Homologous temperature of some materials.

Latent Heat

Latent heat is equal to the amount of energy necessary for the melting of unit mass of crystal material at the freezing point. Materials with high latent heat are more stable.

Table: Latent heat of some materials.

Material	Latentc heat, J/kg
Ice	334000
Lead	23100
Copper	214000
Iron	270000
Mercury	11800

Thermal Expansion

When heat is passed through a material, its shape changes. Generally, a material expands when heated. This property of a material is called Thermal Expansion. There can be change in area, volume and shape of the material.

For example, railway tracks often expand and as a result, get misshapen due to extreme heat.

Thermal Stress

The stress experienced by a body due to either thermal expansion or contraction is called thermal stress. It can be potentially destructive in nature as it can make the material explode.

For example, cracks can be seen on roads where the heat is extreme. The crack is a result of thermal stress.

Engineering Properties

Engineering or more often called mechanical properties deals with material's response to tension, compression, shear, and repeated combined actions of these three. Some engineering concepts are required for understanding these properties, and it is helpful for a general reader to acquire some of these concepts in order to understand the material world. On the other hand, engineers have to consider physical and chemical properties in their designs and applications.

Hardness

The Mohs Scale of Hardness:

1. Talc
2. Gypsum
3. Calcite
4. Fluorite
5. Aptite
6. Orthoclase
7. Quartz
8. Topaz

9. Corundum
10. Diamond

Ability to abrade or indent one another is referred to as hardness. There are several scales of hardness, and the oldest one is a Mohs hardness scale, which assigned a number consecutively from 1 to 10 for talc, gypsum, calcite, fluorite, aptite, orthoclase, quartz, topaz, corundum and diamond, respectively. Diamond is the hardest of all substances. This scale is not linear, but relative. The hardness is assigned by a scratching test. A modified Mohs scale gives 15 grade of hardness between talc and diamond, whereas the Knoop values of hardness range from 32 for gypsum to 7000 for diamond. Hardness should be prime considerations in all tools and machine parts. Harder substances are used for cutting tools. Due to their hardness, the most important. applications of small synthetic diamonds are in cutting tools. In these applications, their colors and sizes are not important.

Tensile Characteristics

Hooke's Law states that the strain produced in an elastic body is proportional to the stress which causes it. The stress may be in the form of tensile or shear.

A tensile stress pulls the two ends of a rod or cable apart. When a small tensile stress is applied, the rod or cable lengthens elastically and returns to its original length after the tensile stress is removed. The ratio of tensile stress to lengthening (strain) divided by the area of the cross section is called modulus of elasticity. The (maximum) tensile stress or load causing permanent or inelastic deformation is called yield load. At the yield load, the lengthening will continue without adding more tensile stress. If the yield load is not removed, the rod lengthening continues until it breaks. The tensile strength is the yield load divided by area of the cross section of a rod or bar. Thus, tensile strengths have the units of force per unit area. The yield load of a cable or rod is equal to the tensile strength multiplied by the cross section area.

Shear Characteristics

Shear stress is a pair of force applied in opposite direction but in a sliding fashion. In response to the shear stress, a material deforms. Usually, shear stress causes adjacent laminar elements of a solid body to slide over each other of. In the case of homogeneous isotopic elastic medium, the ratio of shear strain to the shear stress is called shear modulus. Shear strength is the stress, usually expressed in force unit per unit area, required to produce fracture when impressed perpendicular upon the cross-section of a material.

Shear strengths are important for bolts, rivets, drive key, cutting, and polishing applications.

Fatigue Strength

We all experience that repeated bending of a steel wire on the same kink-point results in breaking it. Repeated loading and unloading of a specific stress on a piece of material till it fails is called fatigue strength or endurance limit. The stress can be tensile, compressed, shear, bending, or a combination of these. Fatigue strengths are hard to measure because of the varying and many stress types. Usually some percentage of tensile and compression strength are taken as a possible fatigue strength. Engineering designs allow wide margins for safety to avoid failure due to fatigue.

Appearance and Dimensional Properties

Color and surface finish give an object an appearance, which is a very important factor in engineering and commercial products. Combining color and surface finish gives artistic designs appealing to customers.

In terms of color, we are interested in the property of a material to filter and reflect lights. Our reaction to and perception of lights depends on the wavelengths. For engineering, comparison, and communication purposes, we must have standard measurements and specifications of color. Four parameters, three for the relative amounts of red, green, and blue components and one for the brightness, are required to specify a color.

Specific concepts and terms must be developed for the description of surface features. There are more than 20 mathematical parameters applied to surface description and some of the terms are: roughness, irregular features of wave, height, width, lay, and direction on the surface; camber, deviation from straightness; out of flat, measure of macroscopic deviations from flatness of a surface.

Furthermore, the surface properties not only affect visual perception, they affect engineering applications as well. Rough surfaces in moving contacts leads to wearing, whereas over smoothness is avoided because of high cost for their manufacture. The importance of surface is indicated by the many methods developed for its analysis. Optical or laser technology can be used to measure large surface areas, whereas visual examination detects features of the order of millimeters. Finer surface features are examined by microscopy, which enlarge the surface by a factor of 10 to 3000.

Still finer surface features are examined using scanning electron microscopy (SEM), which enlarges a surface up to 10,000,000 times. Some SEM machines are now equipped with an energy dispersive X-ray analyzer (EDX), which identifies the chemical elements in the bombarded area. The composition of the surface layers is often determined by electron spectroscopy for chemical analysis (ESCA). The surface is bombarded by low-energy X-rays to emit photoelectrons, the analysis of which gives the atomic composition of the surface layers. Another method yielding composition information is called secondary ion mass spectroscopy (SIMS), which bombards the surface with ions to remove some of the atoms. These atoms are ionized and they pass through a mass spectrometer for further analysis. Auger electron spectroscopy (AES) bombards the surface with an electron beam, which causes the ejection of Auger electrons. The Auger electron is chemical element specific, and they reveal also composition at the surface.

Dimensional properties deal with size and shape, which is also related to quantities. In a modern society, material requirement must be very carefully specified regarding the details such as the surface texture, flatness, allowable defects, shape, dimensions, camber, tolerance, etc. A slight misinterpretation or miscommunication causes not only lots of money, but also a lot of frustration.

Chemical Property

It is important to have the knowledge of chemical properties of engineering materials. Because most the engineering materials come into contact with other materials and react chemically to each other. Due to this chemical reaction they may suffer from chemical deterioration. Some of the chemical properties of engineering materials are listed below:

1. Chemical composition
2. Atomic bonding
3. Corrosion resistance
4. Acidity or Alkalinity

Chemical Composition

The chemical composition of engineering material indicates the elements which are combined together to form that material. Chemical composition of a material effects the properties of engineering materials very much. The strength, hardness, ductility, brittleness, corrosion resistance, weld ability etc. depends on chemical composition of materials.

Hence, we should also have the knowledge of chemical composition of engineering materials. For Example the Chemical compositions of some materials are listed below:

Sl. No.	Material	Chemical Composition
1.	Steel	Fe, Cr, Ni
2.	Brass	Cu = 90%, Ni = 10%
3.	Bronze	90% Cu, 10% Ni
4.	Invar	Fe = 64%, Ni = 36%
5.	Gun Metal	Cu = 88%, Tin = 10%, Zn = 2%
6.	German Silver or Nickel Silver or Electrum	Cu = 50%, Zn = 30%, Ni = 20%
7.	Nichrome	Ni = 60%, Cr = 15%, Fe = 25%
8.	Phosphor Bronge	Cu = 89 – 95.50% , Tin = 3.50 -10%, P = 1%
9.	Manganin	Cu = 84%, Mn = 12%, Ni = 4%
10.	Constantan	Cu = 60%, Ni = 40%

Atomic Bonding

Atomic bonding represents how atoms are bounded to each other to form the material. Many properties, such as melting point, boiling point, thermal conductivity and electrical conductivity of materials are governed by atomic bonding of materials. Hence, to understand the properties of materials, it is very important to study the atomic bonding of materials. Atomic bonds in materials are of following types:

1. Ionic bond: froms by exchanging of valence electrons between atoms.
2. Covalent bonds: froms by sharing of electrons between atoms.
3. Metallic bonds: found in metals.

Corrosion Resistance

Corrosion is a gradual chemical or electromechemical attack on a metal by its surrounding medium. Due to the corrosion, metal starts getting converted into an oxide, salt or some other compound. Corrosion of a metal is effected by many factors such as air, industrial atmosphere, acid, bases, slat solutions and soils etc. Corrosion has a very adverse effect on materials. Due to corrosion, the strength and life of a material is reduced.

Corrosion resistance of a material is the ability of material to resist the oxidation in atmospheric condition. Generally pure metals such as iron, copper, aluminum etc. gets corroded in slowly in atmosphere. To avoid the corrosion of these metal in pure form, we use these metals in the form of alloys such as stainless steel, brass, bronze, German silver, Gunmetal etc.

Acidity or Alkalinity

Acidity or Alkalinity is an important chemical property of engineering materials. A material is acetic or Alkane, it is decided by the ph value of the material. Ph value of a material varies from 0 to 14. Ph value of 7 is considered to be neutral. Ordinary water is having ph value of 7. The materials which are having ph value below 7 are called Acetic and Materials which are having ph value greater than 7 are called alkane. Acidity of Alkalinity of material indicates that how they react with other materials.

References

- Thermal-properties-of-materials, physics: byjus.com, Retrieved 14, March 2020
- Chemical-properties-of-materials: electrical4u.com, Retrieved 28, June 2020
- Mechanical-properties-of-engineering-materials: electrical4u.com, Retrieved 23, August 2020
- Electrical, Physical-Chemical, Education-Resources, Community-College, Materials: ndeed.org, Retrieved 13, July 2020
- Propertyp, applychem: science.uwaterloo.ca, Retrieved 08, April 2020

4

Chemical Bonding

Chemical bond is a physical phenomenon in which the chemical substances are held together by attraction of atoms towards each other either by sharing or exchanging of electrons. There are different types of chemical bonds like covalent bond, ionic bond, coordinate covalent bond or dipolar bond etc. This chapter discusses in detail the different types of bonds and chemical bonding.

Chemical Bond

A chemical bond is a lasting attraction between atoms that enables the formation of chemical compounds. The bond may result from the electrostatic force of attraction between atoms with opposite charges, or through the sharing of electrons as in the covalent bonds. The strength of chemical bonds varies considerably; there are "strong bonds" such as covalent or ionic bonds and "weak bonds" such as Dipole-dipole interaction, the London dispersion force and hydrogen bonding.

Since opposite charges attract via a simple electromagnetic force, the negatively charged electrons that are orbiting the nucleus and the positively charged protons in the nucleus attract each other. An electron positioned between two nuclei will be attracted to both of them, and the nuclei will be attracted toward electrons in this position. This attraction constitutes the chemical bond. Due to the matter wave nature of electrons and their smaller mass, they must occupy a much larger amount of volume compared with the nuclei, and this volume occupied by the electrons keeps the atomic nuclei relatively far apart, as compared with the size of the nuclei themselves. This phenomenon limits the distance between nuclei and atoms in a bond.

In general, strong chemical bonding is associated with the sharing or transfer of electrons between the participating atoms. The atoms in molecules, crystals, metals and diatomic gases—indeed most of the physical environment around us—are held together by chemical bonds, which dictate the structure and the bulk properties of matter.

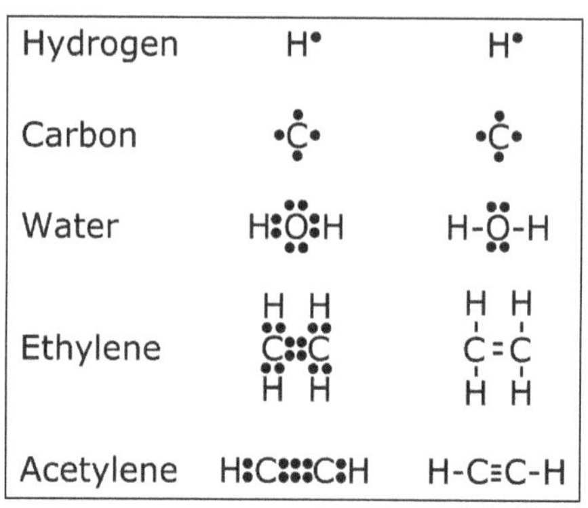

Examples of Lewis dot-style representations of chemical bonds between carbon (C), hydrogen (H), and oxygen (O). Lewis dot diagrams were an early attempt to describe chemical bonding and are still widely used today.

All bonds can be explained by quantum theory, but, in practice, simplification rules allow chemists to predict the strength, directionality, and polarity of bonds. The octet rule and VSEPR theory are two examples. More sophisticated theories are valence bond theory which includes orbital hybridization and resonance, and molecular orbital theory which includes linear combination of atomic orbitals and ligand field theory. Electrostatics are used to describe bond polarities and the effects they have on chemical substances.

Overview of Main Types of Chemical Bonds

A chemical bond is an attraction between atoms. This attraction may be seen as the result of different behaviors of the outermost or valence electrons of atoms. These behaviors merge into each other seamlessly in various circumstances, so that there is no clear line to be drawn between them. However it remains useful and customary to differentiate between different types of bond, which result in different properties of condensed matter.

In the simplest view of a covalent bond, one or more electrons (often a pair of electrons) are drawn into the space between the two atomic nuclei. Energy is released by bond formation. This is not as a reduction in potential energy, because the attraction of the two electrons to the two protons is offset by the electron-electron and proton-proton repulsions. Instead, the release of energy (and hence stability of the bond) arises from the reduction in kinetic energy due to the electrons being in a more spatially distributed (i.e. longer de Broglie wavelength) orbital compared with each electron being confined closer to its respective nucleus. These bonds exist between two particular identifiable atoms and have a direction in space, allowing them to be shown as single connecting lines between atoms in drawings, or modeled as sticks between spheres in models.

In a polar covalent bond, one or more electrons are unequally shared between two nuclei. Covalent bonds often result in the formation of small collections of better-connected atoms called molecules, which in solids and liquids are bound to other molecules by forces that are often much weaker than the covalent bonds that hold the molecules internally together. Such weak intermolecular bonds give organic molecular substances, such as waxes and oils, their soft bulk character, and their low melting points (in liquids, molecules must cease most structured or oriented contact with each other). When covalent bonds link long chains of atoms in large molecules, however (as in polymers such as nylon), or when covalent bonds extend in networks through solids that are not composed of discrete molecules (such as diamond or quartz or the silicate minerals in many types of rock) then the structures that result may be both strong and tough, at least in the direction oriented correctly with networks of covalent bonds. Also, the melting points of such covalent polymers and networks increase greatly.

In a simplified view of an *ionic* bond, the bonding electron is not shared at all, but transferred. In this type of bond, the outer atomic orbital of one atom has a vacancy which allows the addition of one or more electrons. These newly added electrons potentially occupy a lower energy-state (effectively closer to more nuclear charge) than they experience in a different atom. Thus, one nucleus offers a more tightly bound position to an electron than does another nucleus, with the result that one atom may transfer an electron to the other. This transfer causes one atom to assume a net positive charge, and the other to assume a net negative charge. The *bond* then results from electrostatic attraction between atoms and the atoms become positive or negatively charged ions. Ionic bonds may be seen as extreme examples of polarization in covalent bonds. Often, such bonds have no particular orientation in space, since they result from equal electrostatic attraction of each ion to all ions around them. Ionic bonds are strong (and thus ionic substances require high temperatures to melt) but also brittle, since the forces between ions are short-range and do not easily bridge cracks and fractures. This type of bond gives rise to the physical characteristics of crystals of classic mineral salts, such as table salt.

A less often mentioned type of bonding is *metallic* bonding. In this type of bonding, each atom in a metal donates one or more electrons to a "sea" of electrons that reside between many metal atoms. In this sea, each electron is free (by virtue of its wave nature) to be associated with a great many atoms at once. The bond results because the metal atoms become somewhat positively charged due to loss of their electrons while the electrons remain attracted to many atoms, without being part of any given atom. Metallic bonding may be seen as an extreme example of delocalization of electrons over a large system of covalent bonds, in which every atom participates. This type of bonding is often very strong (resulting in the tensile strength of metals). However, metallic bonding is more collective in nature than other types, and so they allow metal crystals to more easily deform, because they are composed of atoms attracted to each other, but not in any particularly-oriented ways. This results in the malleability of metals. The sea of electrons in metallic bonding causes the characteristically good electrical and thermal conductivity of metals, and also their "shiny" reflection of most frequencies of white light.

Early speculations about the nature of the chemical bond, from as early as the 12th century, supposed that certain types of chemical species were joined by a type of chemical affinity. In 1704, Sir Isaac Newton famously outlined his atomic bonding theory, in "Query 31" of his *Opticks*, whereby atoms attach to each other by some "force". Specifically, after acknowledging the various popular theories in vogue at the time, of how atoms were reasoned to attach to each other, i.e. "hooked atoms", "glued together by rest", or "stuck together by conspiring motions", Newton states that he would rather infer from their cohesion, that "particles attract one another by some force, which in immediate contact is exceedingly strong, at small distances performs the chemical operations, and reaches not far from the particles with any sensible effect."

In 1819, on the heels of the invention of the voltaic pile, Jöns Jakob Berzelius developed a theory of chemical combination stressing the electronegative and electropositive characters of the combining atoms. By the mid 19th century, Edward Frankland, F.A. Kekulé, A.S. Couper, Alexander Butlerov, and Hermann Kolbe, building on the theory of radicals, developed the theory of valency, originally called "combining power", in which compounds were joined owing to an attraction of positive and negative poles. In 1916, chemist Gilbert N. Lewis developed the concept of the electron-pair bond, in which two atoms may share one to six electrons, thus forming the single electron bond, a single bond, a double bond, or a triple bond; in Lewis's own words, "An electron may form a part of the shell of two different atoms and cannot be said to belong to either one exclusively."

That same year, Walther Kossel put forward a theory similar to Lewis' only his model assumed complete transfers of electrons between atoms, and was thus a model of ionic bonding. Both Lewis and Kossel structured their bonding models on that of Abegg's rule (1904).

In 1927, the first mathematically complete quantum description of a simple chemical bond, i.e. that produced by one electron in the hydrogen molecular ion, H_2^+, was derived by the Danish physicist Oyvind Burrau. This work showed that the quantum approach to chemical bonds could be fundamentally and quantitatively correct, but the mathematical methods used could not be extended to molecules containing more than one electron. A more practical, albeit less quantitative, approach was put forward in the same year by Walter Heitler and Fritz London. The Heitler-London method forms the basis of what is now called valence bond theory. In 1929, the linear combination of atomic orbitals molecular orbital method (LCAO) approximation was introduced by Sir John Lennard-Jones, who also suggested methods to derive electronic structures of molecules of F_2 (fluorine) and O_2 (oxygen) molecules, from basic quantum principles. This molecular orbital theory represented a covalent bond as an orbital formed by combining the quantum mechanical Schrödinger atomic orbitals which had been hypothesized for electrons in single atoms. The equations for bonding electrons in multi-electron atoms could not be solved to mathematical perfection (i.e., *analytically*), but approximations for them still gave many good qualitative predictions and results. Most quantitative calculations in modern quantum chemistry use either valence bond or molecular orbital theory as a starting point, although a third approach, density functional theory, has become increasingly popular in recent years.

In 1933, H. H. James and A. S. Coolidge carried out a calculation on the dihydrogen molecule that, unlike all previous calculation which used functions only of the distance of the electron from the atomic nucleus, used functions which also explicitly added the distance between the two electrons. With up to 13 adjustable parameters they obtained a result very close to the experimental result for the dissociation energy. Later extensions have used up to 54 parameters and gave excellent agreement with experiments. This calculation convinced the scientific community that quantum theory could give agreement with experiment. However this approach has none of the physical pictures of the valence bond and molecular orbital theories and is difficult to extend to larger molecules.

Bonds in Chemical Formulas

Because atoms and molecules are three-dimensional, it is difficult to use a single method to indicate orbitals and bonds. In molecular formulas the chemical bonds (binding orbitals) between atoms are indicated in different ways depending on the type of discussion. Sometimes, some details are neglected. For example, in organic chemistry one is sometimes concerned only with the functional group of the molecule. Thus, the molecular formula of ethanol may be written in conformational form, three-dimensional form, full two-dimensional form (indicating every bond with no three-dimensional directions), compressed two-dimensional form (CH_3–CH_2–OH), by separating the functional group from another part of the molecule (C_2H_5OH), or by its atomic constituents (C_2H_6O), according to what is discussed. Sometimes, even the non-bonding valence shell electrons (with the two-dimensional approximate directions) are marked, e.g. for elemental carbon 'C'. Some chemists may also mark the respective orbitals, e.g. the hypothetical ethene^{-4} anion (\/C=C/\ $^{-4}$) indicating the possibility of bond formation.

Strong Chemical Bonds

Typical bond lengths in pm and bond energies in kJ/mol. Bond lengths can be converted to Å by division by 100 (1 Å = 100 pm).Data taken from University of Waterloo.		
Bond	**Length (pm)**	**Energy (kJ/mol)**

H – Hydrogen		
H–H	74	436
H–O	96	366
H–F	92	568
H–Cl	127	432
C – Carbon		
C–H	109	413
C–C	154	348
C–C=	151	
=C–C≡	147	
=C–C=	148	
C=C	134	614
C≡C	120	839
C–N	147	308
C–O	143	360
C–F	134	488
C–Cl	177	330
N – Nitrogen		
N–H	101	391
N–N	145	170
N≡N	110	945
O – Oxygen		
O–O	148	145
O=O	121	498
F, Cl, Br, I – Halogens		
F–F	142	158
Cl–Cl	199	243
Br–H	141	366
Br–Br	228	193
I–H	161	298
I–I	267	151

Strong chemical bonds are the *intramolecular* forces which hold atoms together in molecules. A strong chemical bond is formed from the transfer or sharing of electrons between atomic centers and relies on the electrostatic attraction between the protons in nuclei and the electrons in the orbitals.

The types of strong bond differ due to the difference in electronegativity of the constituent elements. A large difference in electronegativity leads to more polar (ionic) character in the bond.

Ionic Bonding

Ionic bonding is a type of electrostatic interaction between atoms which have a large electronegativity difference. There is no precise value that distinguishes ionic from covalent bonding, but a difference of electronegativity of over 1.7 is likely to be ionic, and a difference of less than 1.7 is likely to be covalent. Ionic bonding leads to separate positive and negative ions. Ionic charges are commonly between −3e to +3e. Ionic bonding commonly occurs in metal salts such as sodium chloride (table salt). A typical feature of ionic bonds is that the species form into ionic crystals, in which no ion is specifically paired with any single other ion, in a specific directional bond. Rather, each species of ion is surrounded by ions of the opposite charge, and the spacing between it and each of the oppositely charged ions near it, is the same for all surrounding atoms of the same type. It is thus no longer possible to associate an ion with any specific other single ionized atom near it. This is a situation unlike that in covalent crystals, where covalent bonds between specific atoms are still discernible from the shorter distances between them, as measured via such techniques as X-ray diffraction.

Ionic crystals may contain a mixture of covalent and ionic species, as for example salts of complex acids, such as sodium cyanide, NaCN. X-ray diffraction shows that in NaCN, for example, the bonds between sodium cations (Na^+) and the cyanide anions (CN^-) are *ionic*, with no sodium ion associated with any particular cyanide. However, the bonds between C and N atoms in cyanide are of the *covalent* type, making each of the carbon and nitrogen associated with *just one* of its opposite type, to which it is physically much closer than it is to other carbons or nitrogens in a sodium cyanide crystal.

When such crystals are melted into liquids, the ionic bonds are broken first because they are non-directional and allow the charged species to move freely. Similarly, when such salts dissolve into water, the ionic bonds are typically broken by the interaction with water, but the covalent bonds continue to hold. For example, in solution, the cyanide ions, still bound together as single CN^- ions, move independently through the solution, as do sodium ions, as Na^+. In water, charged ions move apart because each of them are more strongly attracted to a number of water molecules, than to each other. The attraction between ions and water molecules in such solutions is due to a type of weak dipole-dipole type chemical bond. In melted ionic compounds, the ions continue to be attracted to each other, but not in any ordered or crystalline way.

Covalent Bond

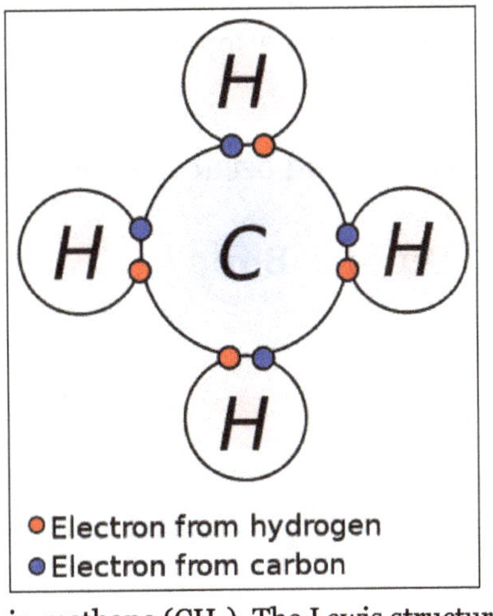

Nonpolar covalent bonds in methane (CH_4). The Lewis structure shows electrons shared between C and H atoms.

Covalent bonding is a common type of bonding, in which two atoms share two valence electrons, one from each of the atoms. In nonpolar covalent bonds, the electronegativity difference between the bonded atoms is small, typically 0 to 0.3. Bonds within most organic compounds are described as covalent. The figure shows methane (CH_4), in which each hydrogen forms a covalent bond with the carbon.

Molecules which are formed primarily from non-polar covalent bonds are often immiscible in water or other polar solvents, but much more soluble in non-polar solvents such as hexane.

A polar covalent bond is a covalent bond with a significant ionic character. This means that the two shared electrons are closer to one of the atoms than the other, creating an imbalance of charge. Such bonds occur between two atoms with moderately different electronegativities and give rise to dipole-dipole interactions. The electronegativity difference between the two atoms in these bonds is 0.3 to 1.7.

Single and Multiple Bonds

A single bond between two atoms corresponds to the sharing of one pair of electrons. The electron density of these two bonding electrons is concentrated in the region between the two atoms, which is the defining quality of a sigma bond.

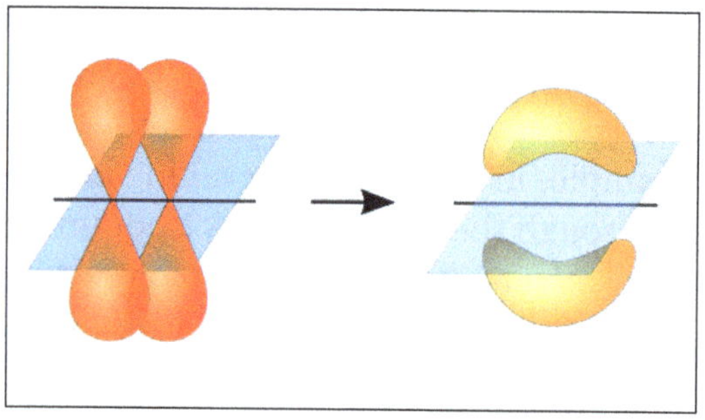

Two p-orbitals forming a pi-bond.

A double bond between two atoms is formed by the sharing of two pairs of electrons, one in a sigma bond and one in a pi bond, with electron density concentrated on two opposite sides of the internuclear axis. A triple bond consists of three shared electron pairs, forming one sigma and two pi bonds.

Quadruple and higher bonds are very rare and occur only between certain transition metal atoms.

Coordinate Covalent Bond (Dipolar Bond)

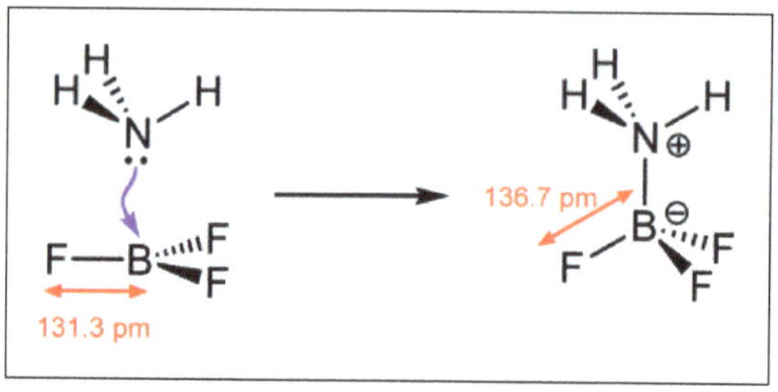

Adduct of ammonia and boron trifluoride.

A coordinate covalent bond is a covalent bond in which the two shared bonding electrons are from the same one of the atoms involved in the bond. For example, boron trifluoride (BF_3) and ammonia (NH_3) from an adduct or coordination complex $F_3B{\leftarrow}NH_3$ with a B–N bond in which a lone pair of electrons on N is shared with an empty atomic orbital on B. BF_3 with an empty orbital is described as an electron pair acceptor or Lewis acid, while NH_3 with a lone pair which can be shared is described as an electron-pair donor or Lewis base. The electrons are shared roughly equally between the atoms in contrast to ionic bonding. Such bonding is shown by an arrow pointing to the Lewis acid.

Transition metal complexes are generally bound by coordinate covalent bonds. For example, the ion Ag^+ reacts as a Lewis acid with two molecules of the Lewis base NH_3 to form the complex ion $Ag(NH_3)_2^+$, which has two Ag←N coordinate covalent bonds.

Metallic Bonding

In metallic bonding, bonding electrons are delocalized over a lattice of atoms. By contrast, in ionic compounds, the locations of the binding electrons and their charges are static. The freely-moving or delocalization of bonding electrons leads to classical metallic properties such as luster (surface light reflectivity), electrical and thermal conductivity, ductility, and high tensile strength.

Intermolecular Bonding

There are four basic types of bonds that can be formed between two or more (otherwise non-associated) molecules, ions or atoms. Intermolecular forces cause molecules to be attracted or repulsed by each other. Often, these define some of the physical characteristics (such as the melting point) of a substance.

- A large difference in electronegativity between two bonded atoms will cause a permanent charge separation, or dipole, in a molecule or ion. Two or more molecules or ions with permanent dipoles can interact within dipole-dipole interactions. The bonding electrons in a molecule or ion will, on average, be closer to the more electronegative atom more frequently than the less electronegative one, giving rise to partial charges on each atom, and causing electrostatic forces between molecules or ions.

- A hydrogen bond is effectively a strong example of an interaction between two permanent dipoles. The large difference in electronegativities between hydrogen and any of fluorine, nitrogen and oxygen, coupled with their lone pairs of electrons cause strong electrostatic forces between molecules. Hydrogen bonds are responsible for the high boiling points of water and ammonia with respect to their heavier analogues.

- The London dispersion force arises due to instantaneous dipoles in neighbouring atoms. As the negative charge of the electron is not uniform around the whole atom, there is always a charge imbalance. This small charge will induce a corresponding dipole in a nearby molecule; causing an attraction between the two. The electron then moves to another part of the electron cloud and the attraction is broken.

- A cation–pi interaction occurs between a pi bond and a cation.

Theories of Chemical Bonding

In the (unrealistic) limit of "pure" ionic bonding, electrons are perfectly localized on one of the two atoms in the bond. Such bonds can be understood by classical physics. The forces between the atoms are characterized by isotropic continuum electrostatic potentials. Their magnitude is in simple proportion to the charge difference.

Covalent bonds are better understood by valence bond theory or molecular orbital theory. The properties of the atoms involved can be understood using concepts such as oxidation number. The electron density within a bond is not assigned to individual atoms, but is instead delocalized between atoms. In valence bond theory, the two electrons on the two atoms are coupled together with the bond strength depending on the overlap between them. In molecular orbital theory, the linear combination of atomic orbitals (LCAO) helps describe the delocalized molecular orbital structures and energies based on the atomic orbitals of the atoms they came from. Unlike pure ionic bonds, covalent bonds may have directed anisotropic properties. These may have their own names, such as sigma bond and pi bond.

In the general case, atoms form bonds that are intermediate between ionic and covalent, depending on the relative electronegativity of the atoms involved. This type of bond is sometimes called polar covalent.

Ionic Bonding

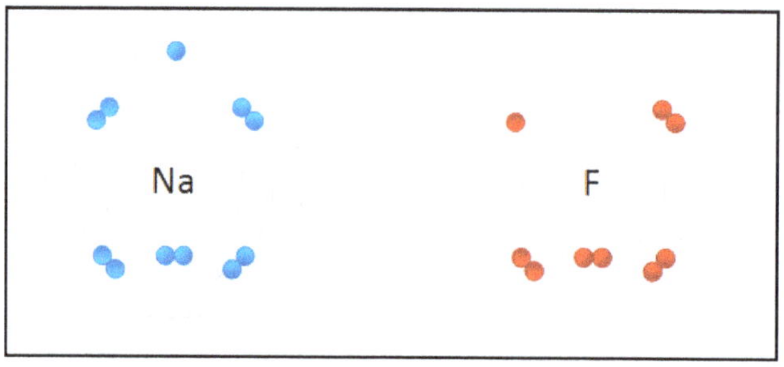

Sodium and fluorine undergoing a redox reaction to form sodium fluoride. Sodium loses its outer electron to give it a stable electron configuration, and this electron enters the fluorine atom exothermically. The oppositely charged ions – typically a great many of them – are then attracted to each other to form a solid.

Ionic bonding is a type of chemical bond that involves the electrostatic attraction between oppositely charged ions, and is the primary interaction occurring in ionic compounds. The ions are atoms that have gained one or more electrons (known as anions, which are negatively charged) and atoms that have lost one or more electrons (known as cations, which are positively charged). This transfer of electrons is known as electrovalence in contrast to covalence. In the simplest case, the cation is a metal atom and the anion is a nonmetal atom, but these ions can be of a more complex nature, e.g. molecular ions like NH_4^+ or SO_4^{2-}. In simpler words, an ionic bond is the transfer of electrons from a metal to a non-metal in order for both atoms to obtain a full valence shell.

It is important to recognize that *clean* ionic bonding – in which one atom or molecule completely share an electron from another – cannot exist: all ionic compounds have some degree of covalent bonding, or electron sharing. Thus, the term "ionic bonding" is given when the ionic character is greater than the covalent character—that is, a bond in which a large electronegativity difference exists between the two atoms, causing the bonding to be more polar (ionic) than in covalent bonding where electrons are shared more equally. Bonds with partially ionic and partially covalent character are called polar covalent bonds.

Ionic compounds conduct electricity when molten or in solution, typically as a solid. Ionic compounds generally have a high melting point, depending on the charge of the ions they consist of. The higher the charges the stronger the cohesive forces and the higher the melting point. They also tend to be soluble in water. Here, the opposite trend roughly holds: the weaker the cohesive forces, the greater the solubility.

Overview

Atoms that have an almost full or almost empty valence shells tend to be very reactive. Atoms that are strongly electronegative (as is the case with halogens) often only have one or two empty orbitals in their valence shell, and frequently bond with other molecules or gain electrons to form anions. Atoms that are weakly electronegative (such as alkali metals) have relatively few valence electrons that can easily be shared with atoms that are strongly electronegative. As a result, weakly electronegative atoms tend to distort their electrons cloud and form cations.

Formation

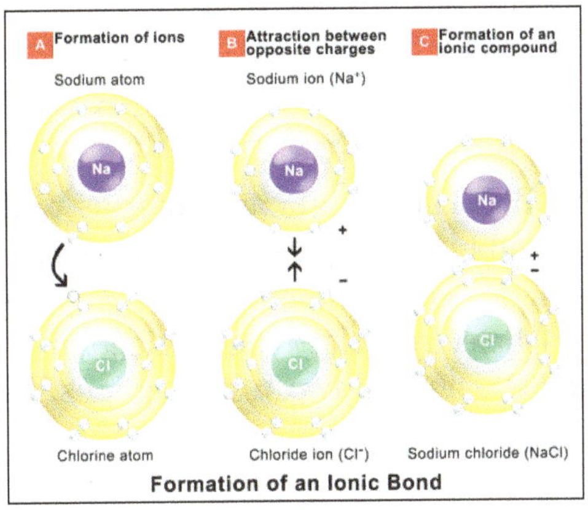

Ionic bonds in sodium chloride.

Ionic bonding can result from a redox reaction when atoms of an element (usually metal), whose ionization energy is low, give some of their electrons to achieve a stable electron configuration. In doing so, cations are formed. The atom of another element (usually nonmetal), whose electron affinity is positive, then accepts the electron(s), again to attain a stable electron configuration, and after accepting electron(s) the atom becomes an anion. Typically, the stable electron configuration is one of the noble gases for elements in the s-block and the p-block, and particular stable electron configurations for d-block and f-block elements. The electrostatic attraction between the anions and cations leads to the formation of a solid with a crystallographic lattice in which the ions are stacked in an alternating fashion. In such a lattice, it is usually not possible to distinguish discrete molecular units, so that the compounds formed are not molecular in nature. However, the ions themselves can be complex and form molecular ions like the acetate anion or the ammonium cation.

For example, common table salt is sodium chloride. When sodium (Na) and chlorine (Cl) are combined, the sodium atoms each lose an electron, forming cations (Na^+), and the chlorine atoms each gain an electron to form anions (Cl^-). These ions are then attracted to each other in a 1:1 ratio to form sodium chloride (NaCl).

$$Na + Cl \rightarrow Na^+ + Cl^- \rightarrow NaCl$$

However, to maintain charge neutrality, strict ratios between anions and cations are observed so that ionic compounds, in general, obey the rules of stoichiometry despite not being molecular compounds. For compounds that are transitional to the alloys and possess mixed ionic and metallic bonding, this may not be the case anymore. Many sulfides, e.g., do form non-stoichiometric compounds.

Many ionic compounds are referred to as salts as they can also be formed by the neutralization reaction of an Arrhenius base like NaOH with an Arrhenius acid like HCl

$$NaOH + HCl \rightarrow NaCl + H_2O$$

The salt NaCl is then said to consist of the acid rest Cl^- and the base rest Na^+.

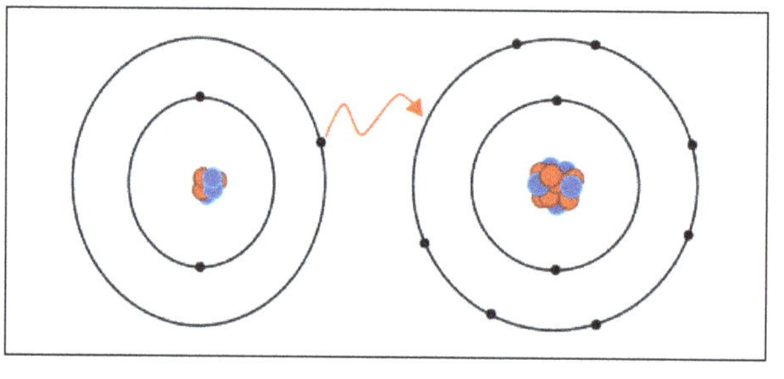

Representation of ionic bonding between lithium and fluorine to form lithium fluoride. Lithium has a low ionization energy and readily gives up its lone valence electron to a fluorine atom, which has a positive electron affinity and accepts the electron that was donated by the lithium atom. The end-result is that lithium is isoelectronic with helium and fluorine is isoelectronic with neon. Electrostatic interaction occurs between the two resulting ions, but typically aggregation is not limited to two of them. Instead, aggregation into a whole lattice held together by ionic bonding is the result.

The removal of electrons from the cation is endothermic, raising the system's overall energy. There may also be energy changes associated with breaking of existing bonds or the addition of more than one electron to form anions. However, the action of the anion's accepting the cation's valence electrons and the subsequent attraction of the ions to each other releases (lattice) energy and, thus, lowers the overall energy of the system.

Ionic bonding will occur only if the overall energy change for the reaction is favorable. In general, the reaction is exothermic, but, e.g., the formation of mercuric oxide (HgO) is endothermic. The charge of the resulting ions is a major factor in the strength of ionic bonding, e.g. a salt C^+A^- is held together by electrostatic forces roughly four times weaker than $C^{2+}A^{2-}$ according to Coulombs law, where C and A represent a generic cation and anion respectively. Of course the sizes of the ions and the particular packing of the lattice are ignored in this simple argument.

Structures

Ionic compounds in the solid state form lattice structures. The two principal factors in determining the form of the lattice are the relative charges of the ions and their relative sizes. Some structures are adopted by a number of compounds; for example, the structure of the rock salt sodium chloride is also adopted by many alkali halides, and binary oxides such as MgO. Pauling's rules provide guidelines for predicting and rationalizing the crystal structures of ionic crystals.

Bond Strength

For a solid crystalline ionic compound the enthalpy change in forming the solid from gaseous ions is termed the lattice energy. The experimental value for the lattice energy can be determined using the Born-Haber cycle. It can also be calculated (predicted) using the Born-Landé equation as the sum of the electrostatic potential energy, calculated by summing interactions between cations and anions, and a short-range repulsive potential energy term. The electrostatic potential can be expressed in terms of the inter-ionic separation and a constant (Madelung constant) that takes account of the geometry of the crystal. The further away from the nucleus the weaker the shield. The Born-Landé equation gives a reasonable fit to the lattice energy of, e.g., sodium chloride, where the calculated (predicted) value is −756 kJ/mol, which compares to −787 kJ/mol using the Born-Haber cycle.

Polarization Effects

Ions in crystal lattices of purely ionic compounds are spherical; however, if the positive ion is small and/or highly charged, it will distort the electron cloud of the negative ion, an effect summarised in Fajans' rules. This polarization of the negative ion leads to a build-up of extra charge density between the two nuclei, i.e., to partial covalency. Larger negative ions are more easily polarized, but the effect is usually important only when positive ions with charges of 3+ (e.g., Al^{3+}) are involved. However, 2+ ions (Be^{2+}) or even 1+ (Li^{+}) show some polarizing power because their sizes are so small (e.g., LiI is ionic but has some covalent bonding present). Note that this is not the ionic polarization effect that refers to displacement of ions in the lattice due to the application of an electric field.

Comparison with Covalent Bonding

In ionic bonding, the atoms are bound by attraction of opposite ions, whereas, in covalent bonding, atoms are bound by sharing electrons to attain stable electron configurations. In covalent bonding, the molecular geometry around each atom is determined by valence shell electron pair repulsion VSEPR rules, whereas, in ionic materials, the geometry follows maximum packing rules. One could say that covalent bonding is more *directional* in the sense that the energy penalty for not adhering to the optimum bond angles is large, whereas ionic bonding has no such penalty. There are no shared electron pairs to repel each other, the ions should simply be packed as efficiently as possible. This often leads to much higher coordination numbers. In NaCl, each ion has 6 bonds and all bond angles are 90 degrees. In CsCl the coordination number is 8. By comparison carbon typically has a maximum of four bonds.

Purely ionic bonding cannot exist, as the proximity of the entities involved in the bonding allows some degree of sharing electron density between them. Therefore, all ionic bonding has some covalent character. Thus, bonding is considered ionic where the ionic character is greater than the covalent character. The larger the difference in electronegativity between the two types of atoms involved in the bonding, the more ionic (polar) it is. Bonds with partially ionic and partially covalent character are called polar covalent bonds. For example, Na–Cl and Mg–O interactions have a few percent covalency, while Si–O bonds are usually ~50% ionic and ~50% covalent. Pauling estimated that an electronegativity difference of 1.7 (on the Pauling scale) corresponds to 50% ionic character, so that a difference greater than 50% corresponds to a bond which is predominantly ionic. Ionic character in covalent bonds can be directly measured for atoms having quadrupolar nuclei (^{2}H, ^{14}N, ^{81,79}Br, ^{35,37}Cl or ^{127}I). These nuclei are generally objects of NQR nuclear quadrupole resonance and NMR nuclear magnetic resonance studies. Interactions between the nuclear quadrupole moments Q and the electric field gradients (EFG) are characterized via the nuclear quadrupole coupling constants QCC = $e^2q_{zz}Q/h$ where the eq_{zz} term corresponds to the principal component of the EFG tensor and e is the elementary charge. In turn, the electric field gradient opens the way to description of bonding modes in molecules when the QCC values are accurately determined by NMR or NQR methods.

In general, when ionic bonding occurs in the solid (or liquid) state, it is not possible to talk about a single "ionic bond" between two individual atoms, because the cohesive forces that keep the lattice together are of a more collective nature. This is quite different in the case of covalent bonding, where we can often speak of a distinct bond localized between two particular atoms. However, even if ionic bonding is combined with some covalency, the result is *not* necessarily discrete bonds of a localized character. In such cases, the resulting bonding often requires description in terms of a band structure consisting of gigantic molecular orbitals spanning the entire crystal. Thus, the bonding in the solid often retains its collective rather than localized nature. When the difference in electronegativity is decreased, the bonding may then lead to a semiconductor, a semimetal or eventually a metallic conductor with metallic bonding.

Covalent Bond

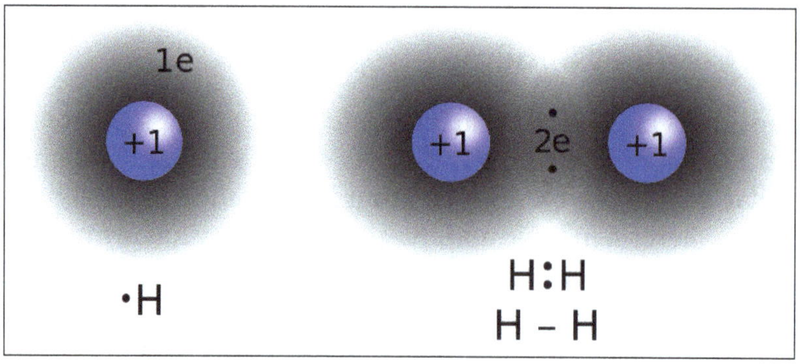

A covalent bond forming H_2 (right) where two hydrogen atoms share the two electrons.

A covalent bond, also called a molecular bond, is a chemical bond that involves the sharing of electron pairs between atoms. These electron pairs are known as shared pairs or bonding pairs, and the stable balance of attractive and repulsive forces between atoms, when they share electrons, is known as covalent bonding. For many molecules, the sharing of electrons allows each atom to attain the equivalent of a full outer shell, corresponding to a stable electronic configuration.

Covalent bonding includes many kinds of interactions, including σ-bonding, π-bonding, metal-to-metal bonding, agostic interactions, bent bonds, and three-center two-electron bonds. The term *covalent bond* dates from 1939. The prefix *co-* means *jointly, associated in action, partnered to a lesser degree,* etc.; thus a "co-valent bond", in essence, means that the atoms share "valence", such as is discussed in valence bond theory.

In the molecule H2, the hydrogen atoms share the two electrons via covalent bonding. Covalency is greatest between atoms of similar electronegativities. Thus, covalent bonding does not necessarily require that the two atoms be of the same elements, only that they be of comparable electronegativity. Covalent bonding that entails sharing of electrons over more than two atoms is said to be delocalized.

History

The term *covalence* in regard to bonding was first used in 1919 by Irving Langmuir in a *Journal of the American Chemical Society* article entitled "The Arrangement of Electrons in Atoms and Molecules". Langmuir wrote that "we shall denote by the term *covalence* the number of pairs of electrons that a given atom shares with its neighbors."

The idea of covalent bonding can be traced several years before 1919 to Gilbert N. Lewis, who in 1916 described the sharing of electron pairs between atoms. He introduced the *Lewis notation* or *electron dot notation* or *Lewis dot structure,* in which valence electrons (those in the outer shell) are represented as dots around the atomic symbols. Pairs of electrons located between atoms represent covalent bonds. Multiple pairs represent multiple bonds, such as double bonds and triple bonds. An alternative form of representation, not shown here, has bond-forming electron pairs represented as solid lines.

Lewis proposed that an atom forms enough covalent bonds to form a full (or closed) outer electron shell. In the methane diagram shown here, the carbon atom has a valence of four and is, therefore, surrounded by eight electrons (the octet rule), four from the carbon itself and four from the hydrogens bonded to it. Each hydrogen has a valence of one and is surrounded by two electrons (a duet rule) – its own one electron plus one from the carbon. The numbers of electrons correspond to full shells in the quantum theory of the atom; the outer shell of a carbon atom is the $n = 2$ shell, which can hold eight electrons, whereas the outer (and only) shell of a hydrogen atom is the $n = 1$ shell, which can hold only two.

While the idea of shared electron pairs provides an effective qualitative picture of covalent bonding, quantum mechanics is needed to understand the nature of these bonds and predict the structures and properties of simple molecules. Walter Heitler and Fritz London are credited with the first successful quantum mechanical explanation of a chemical bond (molecular hydrogen) in 1927. Their work was based on the valence bond model, which assumes that a chemical bond is formed when there is good overlap between the atomic orbitals of participating atoms.

Types of Covalent Bonds

Atomic orbitals (except for s orbitals) have specific directional properties leading to different types of covalent bonds. Sigma (σ) bonds are the strongest covalent bonds and are due to head-on overlapping of orbitals on two different atoms. A single bond is usually a σ bond. Pi (π) bonds are weaker and are due to lateral overlap between p (or d) orbitals. A double bond between two given atoms consists of one σ and one π bond, and a triple bond is one σ and two π bonds.

Covalent bonds are also affected by the electronegativity of the connected atoms which determines the chemical polarity of the bond. Two atoms with equal electronegativity will make nonpolar covalent bonds such as H–H. An unequal relationship creates a polar covalent bond such as with H–Cl.

Covalent Structures

There are several types of structures for covalent substances, including individual molecules, molecular structures, macromolecular structures and giant covalent structures. Individual molecules have strong bonds that hold the atoms together, but there are negligible forces of attraction between molecules. Such covalent substances are usually gases, for example, HCl, SO_2, CO_2, and CH_4. In molecular structures, there are weak forces of attraction. Such covalent substances are low-boiling-temperature liquids (such as ethanol), and low-melting-temperature solids (such as iodine and solid CO_2). Macromolecular structures have large numbers of atoms linked by covalent bonds in chains, including synthetic polymers such as polyethylene and nylon, and biopolymers such as proteins and starch. Network covalent structures (or giant covalent structures) contain large numbers of atoms linked in sheets (such as graphite), or 3-dimensional structures (such as diamond and quartz). These substances have high melting and boiling points, are frequently brittle, and tend to have high electrical resistivity. Elements that have high electronegativity, and the ability to form three or four electron pair bonds, often form such large macromolecular structures.

One- and Three-electron Bonds

Bonds with one or three electrons can be found in radical species, which have an odd number of electrons. The simplest example of a 1-electron bond is found in the dihydrogen cation, H+2. One-electron bonds often have about half the bond energy of a 2-electron bond, and are therefore called "half bonds". However, there are exceptions: in the case of dilithium, the bond is actually stronger for the 1-electron Li+2 than for the 2-electron Li_2. This exception can be explained in terms of hybridization and inner-shell effects.

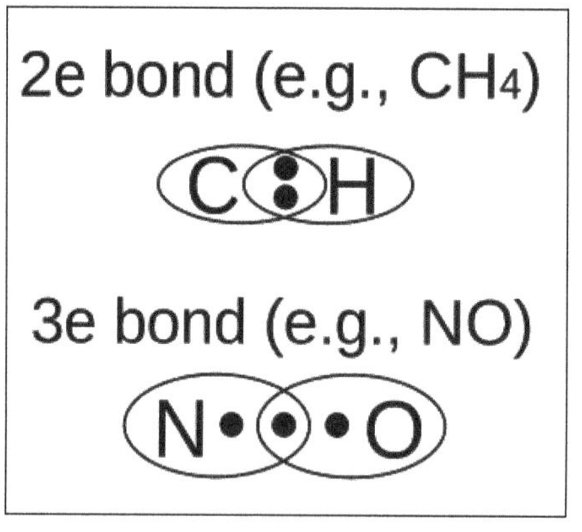

Comparison of the electronic structure of the three-electron bond to the conventional covalent bond.

The simplest example of three-electron bonding can be found in the helium dimer cation, He+2. It is considered a "half bond" because it consists of only one shared electron (rather than two); in molecular orbital terms, the third electron is in an anti-bonding orbital which cancels out half of the bond formed by the other two electrons. Another example of a molecule containing a 3-electron bond, in addition to two 2-electron bonds, is nitric oxide, NO. The oxygen molecule, O_2 can also be regarded as having two 3-electron bonds and one 2-electron bond, which accounts for its paramagnetism and its formal bond order of 2. Chlorine dioxide and its heavier analogues bromine dioxide and iodine dioxide also contain three-electron bonds.

Molecules with odd-electron bonds are usually highly reactive. These types of bond are only stable between atoms with similar electronegativities.

Resonance

There are situations whereby a single Lewis structure is insufficient to explain the electron configuration in a molecule, hence a superposition of structures are needed. The same two atoms in such molecules can be bonded differently in different structures (a single bond in one, a double bond in another, or even none at all), resulting in a non-integer bond order. The nitrate ion is one such example with three equivalent structures. The bond between the nitrogen and each oxygen is a double bond in one structure and a single bond in the other two, so that the average bond order for each N–O interaction is 2 + 1 + 1/3 = 4/3.

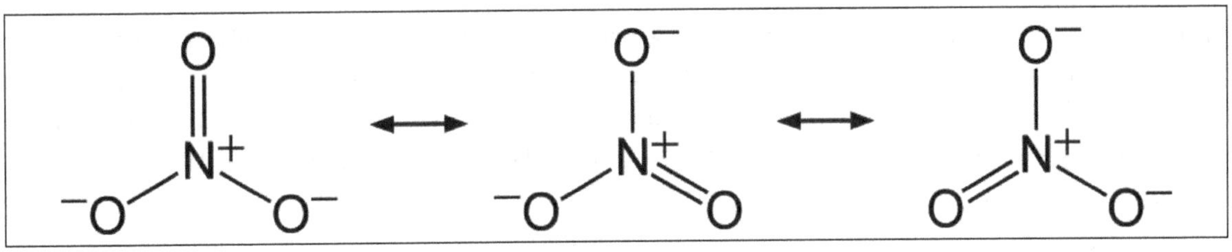

Aromaticity

In organic chemistry, when a molecule with a planar ring obeys Hückel's rule, where the number of π electrons fit the formula $4n + 2$ (where n is an integer), it attains extra stability and symmetry. In benzene, the prototypical aromatic compound, there are 6 π bonding electrons ($n = 1$, $4n + 2 = 6$). These occupy three delocalized π molecular orbitals (molecular orbital theory) or form conjugate π bonds in two resonance structures that linearly combine (valence bond theory), creating a regular hexagon exhibiting a greater stabilization than the hypothetical 1,3,5-cyclohexatriene.

In the case of heterocyclic aromatics and substituted benzenes, the electronegativity differences between different parts of the ring may dominate the chemical behaviour of aromatic ring bonds, which otherwise are equivalent.

Hypervalence

Certain molecules such as xenon difluoride and sulfur hexafluoride have higher co-ordination numbers than would be possible due to strictly covalent bonding according to the octet rule. This is explained by the three-center four-electron bond ("3c–4e") model which interprets the molecular wavefunction in terms of non-bonding highest occupied molecular orbitals in molecular orbital theory and ionic-covalent resonance in valence bond theory.

Electron-deficiency

In three-center two-electron bonds ("3c–2e") three atoms share two electrons in bonding. This type of bonding occurs in electron deficient compounds like diborane. Each such bond (2 per molecule in diborane) contains a pair of electrons which connect the boron atoms to each other in a banana shape, with a proton (nucleus of a hydrogen atom) in the middle of the bond, sharing electrons with both boron atoms. In certain cluster compounds, so-called four-center two-electron bonds also have been postulated.

Quantum Mechanical Description

After the development of quantum mechanics, two basic theories were proposed to provide a quantum description of chemical bonding: valence bond (VB) theory and molecular orbital (MO) theory. A more recent quantum description is given in terms of atomic contributions to the electronic density of states.

Metallic Bonding

Metallic bonding arises from the electrostatic attractive force between conduction electrons (in the form of an electron cloud of delocalized electrons) and positively charged metal ions. It may be described as the sharing of *free* electrons among a lattice of positively charged ions (cations). Metallic bonding accounts for many physical properties of metals, such as strength, ductility, thermal and electrical resistivity and conductivity, opacity, and luster.

Metallic bonding is not the only type of chemical bonding a metal can exhibit, even as a pure substance. For example, elemental gallium consists of covalently-bound pairs of atoms in both liquid and solid state—these pairs form a crystal lattice with metallic bonding between them. Another example of a metal–metal covalent bond is mercurous ion (Hg2+2).

History

As chemistry developed into a science it became clear that metals formed the large majority of the periodic table of the elements and great progress was made in the description of the salts that can be formed in reactions with acids. With the advent of electrochemistry it became clear that metals generally go into solution as positively charged ions and the oxidation reactions of the metals became well understood in the electrochemical series. A picture emerged of metals as positive ions held together by an ocean of negative electrons.

With the advent of quantum mechanics this picture was given more formal interpretation in the form of the free electron model and its further extension, the nearly free electron model. In both of these models the electrons are seen as a gas traveling through the lattice of the solid with an energy that is essentially isotropic in that it depends on the square of the magnitude, *not* the direction of the momentum vector k. In three-dimensional k-space, the set of points of the highest filled levels (the Fermi surface) should therefore be a sphere. In the nearly free correction of the model, box-like Brillouin zones are added to k-space by the periodic potential experienced from the (ionic) lattice, thus mildly breaking the isotropy.

The advent of X-ray diffraction and thermal analysis made it possible to study the structure of crystalline solids, including metals and their alloys, and the construction of phase diagrams became accessible. Despite all this progress the nature of intermetallic compounds and alloys largely remained a mystery and their study was often empirical. Chemists generally steered away from anything that did not seem to follow Dalton's laws of multiple proportions and the problem was considered the domain of a different science, metallurgy.

The almost-free electron model was eagerly taken up by some researchers in this field, notably Hume-Rothery, in an attempt to explain why certain intermetallic alloys with certain compositions would form and others would not. Initially his attempts were quite successful. His idea was to add electrons to inflate the spherical Fermi-balloon inside the series of Brillouin-boxes and determine when a certain box would be full. This indeed predicted a fairly large number of observed alloy compositions. Unfortunately, as soon as cyclotron resonance became available and the shape of the balloon could be determined, it was found that the assumption that the balloon was spherical did not hold at all, except perhaps in the case of caesium. This reduced many of the conclusions to examples of how a model can sometimes give a whole series of correct predictions, yet still be wrong.

The free-electron debacle showed researchers that the model assuming that the ions were in a sea of free electrons needed modification, and so a number of quantum mechanical models such as band structure calculations based on molecular orbitals or the density functional theory were developed. In these models, one either departs from the atomic orbitals of neutral atoms that share their electrons or (in the case of density functional theory) departs from the total electron density. The free-electron picture has, nevertheless, remained a dominant one in education.

The electronic band structure model became a major focus not only for the study of metals but even more so for the study of semiconductors. Together with the electronic states, the vibrational states were also shown to form bands. Rudolf Peierls showed that, in the case of a one-dimensional row of metallic atoms, say hydrogen, an instability had to arise that would lead to the breakup of such a chain into individual molecules. This sparked an interest in the general question: When is collective metallic bonding stable and when will a more localized form of bonding take its place? Much research went into the study of clustering of metal atoms.

As powerful as the concept of the band structure proved to be in the description of metallic bonding, it does have a drawback. It remains a one-electron approximation to a multitudinous many-body problem. In other words, the energy states of each electron are described as if all the other electrons simply form a homogeneous background. Researchers like Mott and Hubbard realized that this was perhaps appropriate for strongly delocalized s- and p-electrons but for d-electrons, and even more for f-electrons the interaction with electrons (and atomic displacements) in the local environment may become stronger than the delocalization that leads to broad bands. Thus, the transition from localized unpaired electrons to itinerant ones partaking in metallic bonding became more comprehensible.

The Nature of Metallic Bonding

The combination of two phenomena gives rise to metallic bonding: delocalization of electrons and the availability of a far larger number of delocalized energy states than of delocalized electrons. The latter could be called electron deficiency.

In 2D

Graphene is an example of two-dimensional metallic bonding. Its metallic bonds are similar to aromatic bonding in benzene, naphthalene, anthracene, ovalene, and so on.

In 3D

Metal aromaticity in metal clusters is another example of delocalization, this time often in three-dimensional entities. Metals take the delocalization principle to its extreme and one could say that a crystal of a metal represents a single molecule over which all conduction electrons are delocalized in all three dimensions. This means that inside the metal one can generally not distinguish molecules, so that the metallic bonding is neither intra- nor intermolecular. 'Nonmolecular' would perhaps be a better term. Metallic bonding is mostly non-polar, because even in alloys there is little difference among the electronegativities of the atoms participating in the bonding interaction (and, in pure elemental metals, none at all). Thus, metallic bonding is an extremely delocalized communal form of covalent bonding. In a sense, metallic bonding is not a 'new' type of bonding at all, therefore, and it describes the bonding only as present in a *chunk* of condensed matter, be it crystalline solid, liquid, or even glass. Metallic vapors by contrast are often atomic (Hg) or at times contain molecules like Na_2 held together by a more conventional covalent bond. This is why it is not correct to speak of a single 'metallic bond'.

The delocalization is most pronounced for s- and p-electrons. For caesium it is so strong that the electrons are virtually free from the caesium atoms to form a gas constrained only by the surface of the metal. For caesium, therefore, the picture of Cs^+ ions held together by a negatively charged electron gas is not too inaccurate. For other elements the electrons are less free, in that they still experience the potential of the metal atoms, sometimes quite strongly. They require a more intricate quantum mechanical treatment (e.g., tight binding) in which the atoms are viewed as neutral, much like the carbon atoms in benzene. For d- and especially f-electrons the delocalization is not strong at all and this explains why these electrons are able to continue behaving as unpaired electrons that retain their spin, adding interesting magnetic properties to these metals.

Electron Deficiency and Mobility

Metal atoms contain few electrons in their valence shells relative to their periods or energy levels. They are electron deficient elements and the communal sharing does not change that. There remain far more available energy states than there are shared electrons. Both requirements for conductivity are therefore fulfilled: strong delocalization and partly filled energy bands. Such electrons can therefore easily change from one energy state into a slightly different one. Thus, not only do they become delocalized, forming a sea of electrons permeating the lattice, but they are also able to migrate through the lattice when an external electrical field is imposed, leading to electrical conductivity. Without the field, there are electrons moving equally in all directions. Under the field, some will adjust their state slightly, adopting a different wave vector. As a consequence, there will be more moving one way than the other and a net current will result.

The freedom of conduction electrons to migrate also give metal atoms, or layers of them, the capacity to slide past each other. Locally, bonds can easily be broken and replaced by new ones after the deformation. This process does not affect the communal metallic bonding very much. This gives rise to metals' typical characteristic phenomena of malleability and ductility. This is particularly true for pure elements. In the presence of dissolved impurities, the defects in the lattice that function as cleavage points may get blocked and the material becomes harder. Gold, for example, is very soft in pure form (24-karat), which is why alloys of 18-karat or lower are preferred in jewelry.

Metals are typically also good conductors of heat, but the conduction electrons only contribute partly to this phenomenon. Collective (i.e., delocalized) vibrations of the atoms known as phonons that travel through the solid as a wave, contribute strongly.

However, the latter also holds for a substance like diamond. It conducts heat quite well but *not* electricity. The latter is *not* a consequence of the fact that delocalization is absent in diamond, but simply that carbon is not electron deficient. The electron deficiency is an important point in distinguishing metallic from more conventional covalent bonding. Thus, we should amend the expression given above into: *Metallic bonding is an extremely delocalized communal form of electron deficient covalent bonding.*

Metallic Radius

Metallic radius is defined as one-half of the distance between the two adjacent metal ions in the metallic lattice. This radius depends on the nature of the atom as well as its environment—specifically, on the coordination number (CN), which in turn depends on the temperature and applied pressure.

When comparing periodic trends in the size of atoms it is often desirable to apply so-called Goldschmidt correction, which converts the radii to the values the atoms would have if they were 12-coordinated. Since metallic radii are always biggest for the highest coordination number, correction for less dense coordinations involves multiplying by x, where $0 < x < 1$. Specifically, for CN = 4, x = 0.88; for CN = 6, x = 0.96, and for CN = 8, x = 0.97. The correction is named after Victor Goldschmidt who obtained the numerical values quoted above.

The radii follow general periodic trends: they decrease across the period due to increase in the effective nuclear charge, which is not offset by the increased number of valence electrons. The radii also increase down the group due to increase in principal quantum number. Between rows 3 and 4, the lanthanide contraction is observed – there is very little increase of the radius down the group due to the presence of poorly shielding f orbitals.

Strength of the Bond

The atoms in metals have a strong attractive force between them. Much energy is required to overcome it. Therefore, metals often have high boiling points, with tungsten (5828 K) being extremely high. A remarkable exception are the elements of the zinc group: Zn, Cd, and Hg. Their electron configuration ends in $...ns^2$ and this comes to resemble a noble gas configuration like that of helium more and more when going down in the periodic table because the energy distance to the empty np orbitals becomes larger. These metals are therefore relatively volatile, and are avoided in ultra-high vacuum systems.

Otherwise, metallic bonding can be very strong, even in molten metals, such as Gallium. Even though gallium will melt from the heat of one's hand just above room temperature, its boiling point is not far from that of copper. Molten gallium is therefore a very nonvolatile liquid thanks to its strong metallic bonding.

The strong bonding of metals in the liquid form demonstrates that the energy of a metallic bond is not a strong function of the direction of the metallic bond; this lack of bond directionality is a direct consequence of electron delocalization, and is best understood in contrast to the directional bonding of covalent bonds. The energy of a metallic bond is thus mostly a function of the amount of electrons which surround the metallic atom, as exemplified by the Embedded atom model. This typically results in metals assuming relatively simple, close-packed crystal structures, such as FCC, BCC, and HCP.

Given high enough cooling rates and appropriate alloy composition, metallic bonding can occur even in glasses with an amorphous structure.

Much biochemistry is mediated by the weak interaction of metal ions and biomolecules. Such interactions and their associated conformational change has been measured using dual polarisation interferometry.

Solubility and Compound Formation

Metals are insoluble in water or organic solvents unless they undergo a reaction with them. Typically this is an oxidation reaction that robs the metal atoms of their itinerant electrons, destroying the metallic bonding. However metals are often readily soluble in each other while retaining the metallic character of their bonding. Gold, for example, dissolves easily in mercury, even at room temperature. Even in solid metals, the solubility can be extensive. If the structures of the two metals are the same, there can even be complete solid solubility, as in the case of electrum, the alloys of silver and gold. At times, however, two metals will form alloys with different structures than either of the two parents. One could call these materials metal compounds, but, because materials with metallic bonding are typically not molecular, Dalton's law of integral proportions is not valid and often a range of stoichiometric ratios can be achieved. It is better to abandon such concepts as 'pure substance' or 'solute' is such cases and speak of phases instead. The study of such phases has traditionally been more the domain of metallurgy than of chemistry, although the two fields overlap considerably.

Localization and Clustering: from Bonding to Bonds

The metallic bonding in complicated compounds does not necessarily involve all constituent elements equally. It is quite possible to have an element or more that do not partake at all. One could picture the conduction electrons flowing around them like a river around an island or a big rock. It is possible to observe which elements do partake, e.g., by looking at the core levels in an X-ray photoelectron spectroscopy (XPS) spectrum. If an element partakes, its peaks tend to be skewed.

Some intermetallic materials e.g. do exhibit metal clusters, reminiscent of molecules and these compounds are more a topic of chemistry than of metallurgy. The formation of the clusters could be seen as a way to 'condense out' (localize) the electron deficient bonding into bonds of a more localized nature. Hydrogen is an extreme example of this form of condensation. At high pressures it is a metal. The core of the planet Jupiter could be said to be held together by a combination of metallic bonding and high pressure induced by gravity. At lower pressures however the bonding becomes entirely localized into a regular covalent bond. The localization is so complete that the (more familiar) H_2 gas results. A similar argument holds for an element like boron. Though it is electron deficient compared to carbon, it does not form a metal. Instead it has a number of complicated structures in which icosahedral B_{12} clusters dominate. Charge density waves are a related phenomenon.

As these phenomena involve the movement of the atoms towards or away from each other, they can be interpreted as the coupling between the electronic and the vibrational states (i.e. the phonons) of the material. A different such electron-phonon interaction is thought to cause a very different result at low temperatures, that of superconductivity. Rather than blocking the mobility of the charge carriers by forming electron pairs in localized bonds, Cooper-pairs are formed that no longer experience any resistance to their mobility.

Optical Properties

The presence of an ocean of mobile charge carriers has profound effects on the optical properties of metals. They can only be understood by considering the electrons as a *collective* rather than considering the states of individual electrons involved in more conventional covalent bonds.

Light consists of a combination of an electrical and a magnetic field. The electrical field is usually able to excite an elastic response from the electrons involved in the metallic bonding. The result is that photons are not able to penetrate very far into the metal and are typically reflected. They bounce off, although some may also be absorbed. This holds equally for all photons of the visible spectrum, which is why metals are often silvery white or grayish with the characteristic specular reflection of metallic luster. The balance between reflection and absorption determines how white or how gray they are, although surface tarnish can obscure such observations. Silver, a very good metal with high conductivity is one of the whitest.

Notable exceptions are reddish copper and yellowish gold. The reason for their color is that there is an upper limit to the frequency of the light that metallic electrons can readily respond to, the plasmon frequency. At the plasmon frequency, the frequency-dependent dielectric function of the free electron gas goes from negative (reflecting) to positive (transmitting); higher frequency photons are not reflected at the surface, and do not contribute to the color of the metal. There are some materials like indium tin oxide (ITO) that are metallic conductors (actually degenerate semiconductors) for which this threshold is in the infrared, which is why they are transparent in the visible, but good mirrors in the IR.

For silver the limiting frequency is in the far UV, but for copper and gold it is closer to the visible. This explains the colors of these two metals. At the surface of a metal resonance effects known as surface plasmons can result. They are collective oscillations of the conduction electrons like a ripple in the electronic ocean. However, even if photons have enough energy they usually do not have enough momentum to set the ripple in motion. Therefore, plasmons are hard to excite on a bulk metal. This is why gold and copper still look like lustrous metals albeit with a dash of color. However, in colloidal gold the metallic bonding is confined to a tiny metallic particle, preventing the oscillation wave of the plasmon from 'running away'. The momentum selection rule is therefore broken, and the plasmon resonance causes an extremely intense absorption in the green with a resulting beautiful purple-red color. Such colors are orders of magnitude more intense than ordinary absorptions seen in dyes and the like that involve individual electrons and their energy states.

Electronegativity

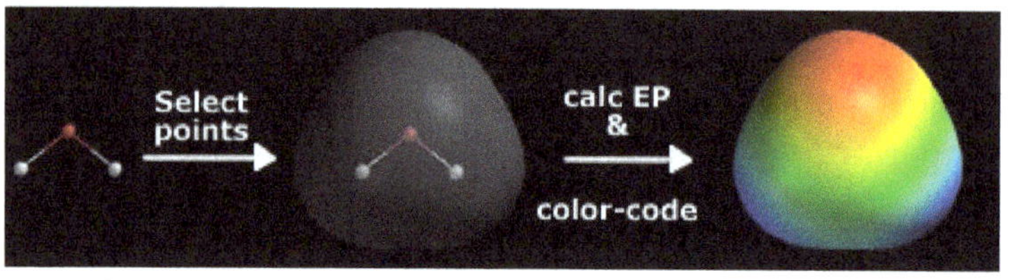

Electrostatic potential map of a water molecule, where the oxygen atom has a more negative charge (red) than the positive (blue) hydrogen atoms.

Electronegativity, symbol χ, is a chemical property that describes the tendency of an atom or a functional group to attract electrons (or electron density) towards itself. An atom's electronegativity is affected by both its atomic number and the distance at which its valence electrons reside from the charged nucleus. The higher the associated electronegativity number, the more an element or compound attracts electrons towards it.

The term "electronegativity" was introduced by Jöns Jacob Berzelius in 1811, though the concept was known even before that and was studied by many chemists including Avogadro. In spite of its long history, an accurate scale of electronegativity had to wait till 1932, when Linus Pauling proposed an electronegativity scale, which depends on bond energies, as a development of valence bond theory. It has been shown to correlate with a number of other chemical properties. Electronegativity cannot be directly measured and must be calculated from other atomic or molecular properties. Several methods of calculation have been proposed, and although there may be small differences in the numerical values of the electronegativity, all methods show the same periodic trends between elements.

The most commonly used method of calculation is that originally proposed by Linus Pauling. This gives a dimensionless quantity, commonly referred to as the Pauling scale, on a relative scale running from around 0.7 to 3.98 (hydrogen = 2.20). When other methods of calculation are used, it is conventional (although not obligatory) to quote the results on a scale that covers the same range of numerical values: this is known as an electronegativity in Pauling units.

As it is usually calculated, electronegativity is not a property of an atom alone, but rather a property of an atom in a molecule. Properties of a free atom include ionization energy and electron affinity. It is to be expected that the electronegativity of an element will vary with its chemical environment, but it is usually considered to be a transferable property, that is to say that similar values will be valid in a variety of situations.

On the most basic level, electronegativity is determined by factors like the nuclear charge (the more protons an atom has, the more "pull" it will have on electrons) and the number/location of other electrons present in the atomic shells (the more electrons an atom has, the farther from the nucleus the valence electrons will be, and as a result the less positive charge they will experience—both because of their increased distance from the nucleus, and because the other electrons in the lower energy core orbitals will act to shield the valence electrons from the positively charged nucleus).

The opposite of electronegativity is electropositivity: a measure of an element's ability to donate electrons.

Caesium is the least electronegative element in the periodic table (=0.79), while fluorine is most electronegative (=3.98). (Francium and caesium were originally both assigned 0.7; caesium's value was later refined to 0.79, but no experimental data allows a similar refinement for francium. However, francium's ionization energy is known to be slightly higher than caesium's, in accordance with the relativistic stabilization of the 7s orbital, and this in turn implies that francium is in fact more electronegative than caesium.)

Electronegativities of the Elements

Periodic table of electronegativity by Pauling scale

→ Atomic radius decreases → Ionization energy increases → Electronegativity increases →

	1	2	3	4	5	6	7	8	9	10	11	12	13	14	15	16	17	18
Group →																		
↓ Period																		
1	H 2.20																	He
2	Li 0.98	Be 1.57											B 2.04	C 2.55	N 3.04	O 3.44	F 3.98	Ne
3	Na 0.93	Mg 1.31											Al 1.61	Si 1.90	P 2.19	S 2.58	Cl 3.16	Ar
4	K 0.82	Ca 1.00	Sc 1.36	Ti 1.54	V 1.63	Cr 1.66	Mn 1.55	Fe 1.83	Co 1.88	Ni 1.91	Cu 1.90	Zn 1.65	Ga 1.81	Ge 2.01	As 2.18	Se 2.55	Br 2.96	Kr 3.00
5	Rb 0.82	Sr 0.95	Y 1.22	Zr 1.33	Nb 1.6	Mo 2.16	Tc 1.9	Ru 2.2	Rh 2.28	Pd 2.20	Ag 1.93	Cd 1.69	In 1.78	Sn 1.96	Sb 2.05	Te 2.1	I 2.66	Xe 2.60
6	Cs 0.79	Ba 0.89	*	Hf 1.3	Ta 1.5	W 2.36	Re 1.9	Os 2.2	Ir 2.20	Pt 2.28	Au 2.54	Hg 2.00	Tl 1.62	Pb 1.87	Bi 2.02	Po 2.0	At 2.2	Rn 2.2
7	Fr 0.7[en 1]	Ra 0.9	**	Rf	Db	Sg	Bh	Hs	Mt	Ds	Rg	Cn	Uut	Fl	Uup	Lv	Uus	Uuo
		*	La 1.1	Ce 1.12	Pr 1.13	Nd 1.14	Pm 1.13	Sm 1.17	Eu 1.2	Gd 1.2	Tb 1.1	Dy 1.22	Ho 1.23	Er 1.24	Tm 1.25	Yb 1.1	Lu 1.27	
		**	Ac 1.1	Th 1.3	Pa 1.5	U 1.38	Np 1.36	Pu 1.28	Am 1.13	Cm 1.28	Bk 1.3	Cf 1.3	Es 1.3	Fm 1.3	Md 1.3	No 1.3	Lr	

Values are given for the elements in their most common and stable oxidation states.

1. Electronegativity of francium was chosen by Pauling as 0.7, close to that of caesium (also assessed 0.7 at that point). The base value of hydrogen was later increased by 0.10 and caesium's electronegativity was later refined to 0.79; however, no refinements have been made for francium as no experiment has been conducted and the old value was kept. However, francium is expected and, to a small extent, observed to be less electropositive than caesium.

Methods of Calculation

Pauling Electronegativity

Pauling first proposed the concept of electronegativity in 1932 as an explanation of the fact that the covalent bond between two different atoms (A–B) is stronger than would be expected by taking the average of the strengths of the A–A and B–B bonds. According to valence bond theory, of which Pauling was a notable proponent, this "additional stabilization" of the heteronuclear bond is due to the contribution of ionic canonical forms to the bonding.

The difference in electronegativity between atoms A and B is given by:

$$|\chi_A - \chi_B| = (\mathrm{eV})^{-1/2}\sqrt{E_d(\mathrm{AB}) - [E_d(\mathrm{AA}) + E_d(\mathrm{BB})]/2}$$

where the dissociation energies, E_d, of the A–B, A–A and B–B bonds are expressed in electronvolts, the factor $(\mathrm{eV})^{-1/2}$ being included to ensure a dimensionless result. Hence, the difference in Pauling electronegativity between hydrogen and bromine is 0.73 (dissociation energies: H–Br, 3.79 eV; H–H, 4.52 eV; Br–Br 2.00 eV).

As only differences in electronegativity are defined, it is necessary to choose an arbitrary reference point in order to construct a scale. Hydrogen was chosen as the reference, as it forms covalent bonds with a large variety of elements: its electronegativity was fixed first at 2.1, later revised to 2.20. It is also necessary to decide which of the two elements is the more electronegative (equivalent to choosing one of the two possible signs for the square root). This is usually done using "chemical intuition": in the above example, hydrogen bromide dissolves in water to form H^+ and Br^- ions, so it may be assumed that bromine is more electronegative than hydrogen. However, in principle, since the same electronegativities should be obtained for any two bonding compounds, the data are in fact overdetermined, and the signs are unique once a reference point is fixed (usually, for H or F).

To calculate Pauling electronegativity for an element, it is necessary to have data on the dissociation energies of at least two types of covalent bond formed by that element. A. L. Allred updated Pauling's original values in 1961 to take account of the greater availability of thermodynamic data, and it is these "revised Pauling" values of the electronegativity that are most often used.

The essential point of Pauling electronegativity is that there is an underlying, quite accurate, semi-empirical formula for dissociation energies, namely:

$$E_d(\mathrm{AB}) = [E_d(\mathrm{AA}) + E_d(\mathrm{BB})]/2 + (\chi_A - \chi_B)^2 eV$$

or sometimes, a more accurate fit,

$$E_d(\mathrm{AB}) = \sqrt{E_d(\mathrm{AA})E_d(\mathrm{BB})} + 1.3(\chi_A - \chi_B)^2 eV$$

This is an approximate equation, but holds with good accuracy. Pauling obtained it by noting that a bond can be approximately represented as a quantum mechanical superposition of a covalent bond and two ionic bond-states. The covalent energy of a bond is approximately, by quantum mechanical calculations, the geometric mean of the two energies of covalent bonds of the same molecules, and there is an additional energy that comes from ionic factors, i.e. polar character of the bond.

The geometric mean is approximately equal to the arithmetic mean - which is applied in the first formula above - when the energies are of the similar value, e.g., except for the highly electropositive elements, where there is a larger difference of two dissociation energies; the geometric mean is more accurate and almost always gives a positive excess energy, due to ionic bonding. The square root of this excess energy, Pauling notes, is approximately additive, and hence one can introduce the electronegativity. Thus, it is this semi-empirical formula for bond energy that underlies Pauling electronegativity concept.

The formulas are approximate, but this rough approximation is in fact relatively good and gives the right intuition, with the notion of polarity of the bond and some theoretical grounding in quantum mechanics. The electronegativities are then determined to best fit the data.

In more complex compounds, there is additional error since electronegativity depends on the molecular environment of an atom. Also, the energy estimate can be only used for single, not for multiple bonds. The energy of formation of a molecule containing only single bonds then can be approximated from an electronegativity table, and depends on the constituents and sum of squares of differences of electronegativities of all pairs of bonded atoms. Such a formula for estimating energy typically has relative error of order of 10%, but can be used to get a rough qualitative idea and understanding of a molecule.

Mulliken Electronegativity

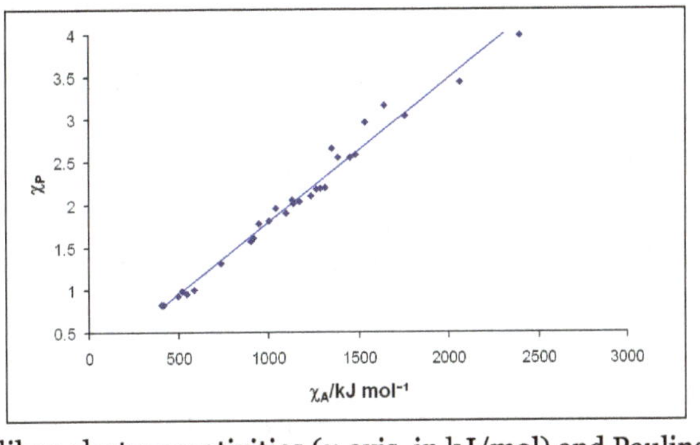

The correlation between Mulliken electronegativities (x-axis, in kJ/mol) and Pauling electronegativities (y-axis).

Robert S. Mulliken proposed that the arithmetic mean of the first ionization energy (E_i) and the electron affinity (E_{ea}) should be a measure of the tendency of an atom to attract electrons. As this definition is not dependent on an arbitrary relative scale, it has also been termed absolute electronegativity, with the units of kilojoules per mole or electronvolts.

$$\chi = (E_i + E_{ea})/2$$

However, it is more usual to use a linear transformation to transform these absolute values into values that resemble the more familiar Pauling values. For ionization energies and electron affinities in electronvolts,

$$\chi = 0.187(E_i + E_{ea}) + 0.17$$

and for energies in kilojoules per mole,

$$\chi = (1.97 \times 10^{-3})(E_i + E_{ea}) + 0.19.$$

The Mulliken electronegativity can only be calculated for an element for which the electron affinity is known, fifty-seven elements as of 2006. The Mulliken electronegativity of an atom is sometimes said to be the negative of the chemical potential. By inserting the energetic definitions of the ionization potential and electron affinity into the Mulliken electronegativity, it is possible to show that the Mulliken chemical potential is a finite difference approximation of the electronic energy with respect to the number of electrons., i.e.,

$$\mu(\text{Mulliken}) = -\chi(\text{Mulliken}) = -(E_i + E_{ea})/2$$

Allred–Rochow Electronegativity

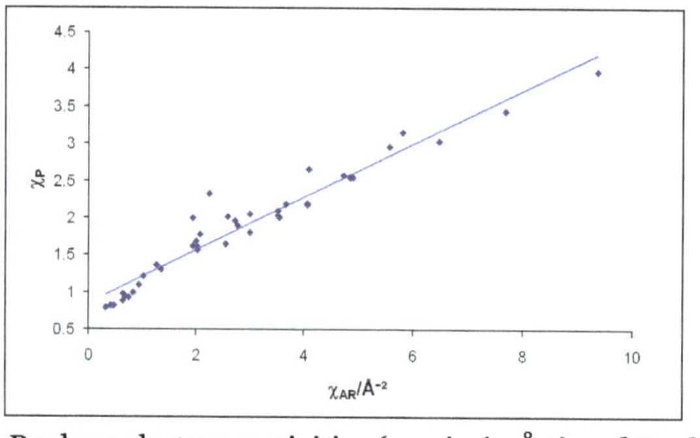

The correlation between Allred–Rochow electronegativities (x-axis, in Å^{-2}) and Pauling electronegativities (y-axis).

A. Louis Allred and Eugene G. Rochow considered that electronegativity should be related to the charge experienced by an electron on the "surface" of an atom: The higher the charge per unit area of atomic surface the greater the tendency of that atom to attract electrons. The effective nuclear charge, Z_{eff}, experienced by valence electrons can be estimated using Slater's rules, while the surface area of an atom in a molecule can be taken to be proportional to the square of the covalent radius, r_{cov}. When r_{cov} is expressed in picometres,

$$\chi = 3590 \frac{Z_{eff}}{r_{cov}^2} + 0.744$$

Sanderson Electronegativity Equalization

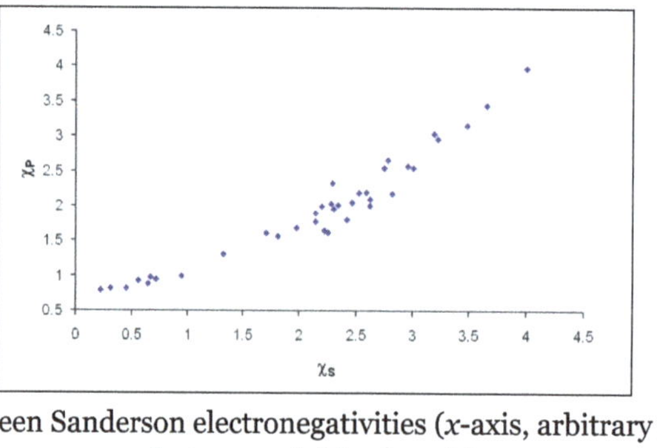

The correlation between Sanderson electronegativities (x-axis, arbitrary units) and Pauling electronegativities (y-axis).

R.T. Sanderson has also noted the relationship between Mulliken electronegativity and atomic size, and has proposed a method of calculation based on the reciprocal of the atomic volume.

With a knowledge of bond lengths, Sanderson's model allows the estimation of bond energies in a wide range of compounds. Sanderson's model has also been used to calculate molecular geometry, *s*-electrons energy, NMR spin-spin constants and other parameters for organic compounds. This work underlies the concept of electronegativity equalization, which suggests that electrons distribute themselves around a molecule to minimize or to equalize the Mulliken electronegativity. This behavior is analogous to the equalization of chemical potential in macroscopic thermodynamics.

Allen Electronegativity

Perhaps the simplest definition of electronegativity is that of Leland C. Allen, who has proposed that it is related to the average energy of the valence electrons in a free atom,

$$\chi = \frac{n_s \varepsilon_s + n_p \varepsilon_p}{n_s + n_p}$$

where $\varepsilon_{s,p}$ are the one-electron energies of s- and p-electrons in the free atom and $n_{s,p}$ are the number of s- and p-electrons in the valence shell. It is usual to apply a scaling factor, 1.75×10^{-3} for energies expressed in kilojoules per mole or 0.169 for energies measured in electronvolts, to give values that are numerically similar to Pauling electronegativities.

The one-electron energies can be determined directly from spectroscopic data, and so electronegativities calculated by this method are sometimes referred to as spectroscopic electronegativities. The necessary data are available for almost all elements, and this method allows the estimation of electronegativities for elements that cannot be treated by the other methods, e.g. francium, which has an Allen electronegativity of 0.67. However, it is not clear what should be considered to be valence electrons for the d- and f-block elements, which leads to an ambiguity for their electronegativities calculated by the Allen method.

In this scale neon has the highest electronegativity of all elements, followed by fluorine, helium, and oxygen.

Correlation of Electronegativity With Other Properties

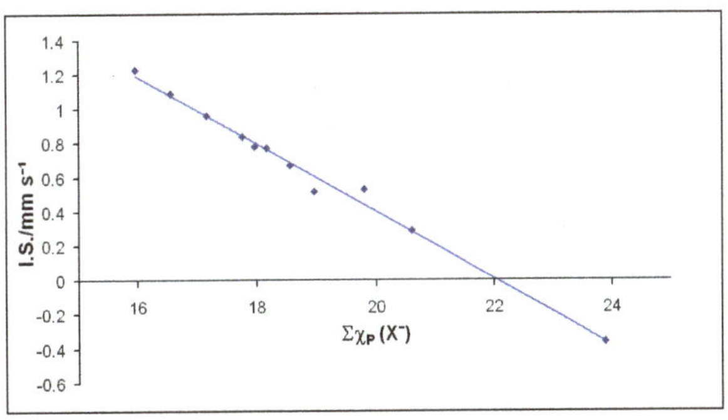

The variation of the isomer shift (*y*-axis, in mm/s) of $[SnX_6]^{2-}$ anions, as measured by ^{119}Sn Mössbauer spectroscopy, against the sum of the Pauling electronegativities of the halide substituents (*x*-axis).

The wide variety of methods of calculation of electronegativities, which all give results that correlate well with one another, is one indication of the number of chemical properties which might be affected by electronegativity. The most obvious application of electronegativities is in the discussion of bond polarity, for which the concept was introduced by Pauling. In general, the greater the difference in electronegativity between two atoms the more polar the bond that will be formed between them, with the atom having the higher electronegativity being at the negative end of the dipole. Pauling proposed an equation to relate "ionic character" of a bond to the difference in electronegativity of the two atoms, although this has fallen somewhat into disuse.

Several correlations have been shown between infrared stretching frequencies of certain bonds and the electronegativities of the atoms involved: however, this is not surprising as such stretching frequencies depend in part on bond strength, which enters into the calculation of Pauling electronegativities. More convincing are the correlations between electronegativity and chemical shifts in NMR spectroscopy or isomer shifts in Mössbauer spectroscopy. Both these measurements depend on the s-electron density at the nucleus, and so are a good indication that the different measures of electronegativity really are describing "the ability of an atom in a molecule to attract electrons to itself".

Trends in Electronegativity

Periodic Trends

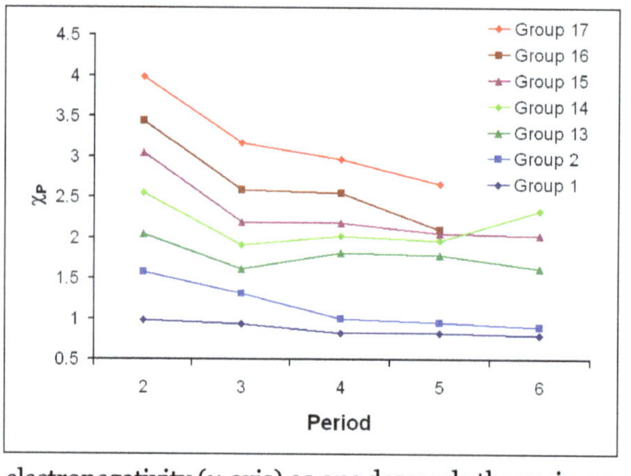

The variation of Pauling electronegativity (y-axis) as one descends the main groups of the periodic table from the second period to the sixth period.

In general, electronegativity increases on passing from left to right along a period, and decreases on descending a group. Hence, fluorine is the most electronegative of the elements (not counting noble gases), whereas caesium is the least electronegative, at least of those elements for which substantial data is available. This would lead one to believe that caesium fluoride is the compound whose bonding features the most ionic character.

There are some exceptions to this general rule. Gallium and germanium have higher electronegativities than aluminium and silicon, respectively, because of the d-block contraction. Elements of the fourth period immediately after the first row of the transition metals have unusually small atomic radii because the 3d-electrons are not effective at shielding the increased nuclear charge, and smaller atomic size correlates with higher electronegativity. The anomalously high electronegativity of lead, in particular when compared to thallium and bismuth, appears to be an artifact of data selection (and data availability)—methods of calculation other than the Pauling method show the normal periodic trends for these elements.

Variation of Electronegativity With Oxidation Number

In inorganic chemistry it is common to consider a single value of the electronegativity to be valid for most "normal" situations. While this approach has the advantage of simplicity, it is clear that the electronegativity of an element is *not* an invariable atomic property and, in particular, increases with the oxidation state of the element.

Allred used the Pauling method to calculate separate electronegativities for different oxidation states of the handful of elements (including tin and lead) for which sufficient data was available. However, for most elements, there are not enough different covalent compounds for which bond dissociation energies are known to make this approach feasible. This is particularly true of the transition elements, where quoted electronegativity values are usually, of necessity, averages over several different oxidation states and where trends in electronegativity are harder to see as a result.

Acid	Formula	Chlorine oxidation state	pK_a
Hypochlorous acid	$HClO$	+1	+7.5
Chlorous acid	$HClO_2$	+3	+2.0
Chloric acid	$HClO_3$	+5	−1.0
Perchloric acid	$HClO_4$	+7	−10

The chemical effects of this increase in electronegativity can be seen both in the structures of oxides and halides and in the acidity of oxides and oxoacids. Hence CrO_3 and Mn_2O_7 are acidic oxides with low melting points, while Cr_2O_3 is amphoteric and Mn_2O_3 is a completely basic oxide.

The effect can also be clearly seen in the dissociation constants of the oxoacids of chlorine. The effect is much larger than could be explained by the negative charge being shared among a larger number of oxygen atoms, which would lead to a difference in pK_a of $\log_{10}(1/4) = -0.6$ between hypochlorous acid and perchloric acid. As the oxidation state of the central chlorine atom increases, more electron density is drawn from the oxygen atoms onto the chlorine, reducing the partial negative charge on the oxygen atoms and increasing the acidity.

Group Electronegativity

In organic chemistry, electronegativity is associated more with different functional groups than with individual atoms. The terms group electronegativity and substituent electronegativity are used synonymously. However, it is common to distinguish between the inductive effect and the resonance effect, which might be described as σ- and π-electronegativities, respectively. There are a number of linear free-energy relationships that have been used to quantify these effects, of which the Hammett equation is the best known. Kabachnik parameters are group electronegativities for use in organophosphorus chemistry.

Electropositivity

Electropositivity is a measure of an element's ability to donate electrons, and therefore form positive ions; thus, it is opposed to electronegativity.

Mainly, this is an attribute of metals, meaning that, in general, the greater the metallic character of an element the greater the electropositivity. Therefore, the alkali metals are most electropositive of all. This is because they have a single electron in their outer shell and, as this is relatively far from the nucleus of the atom, it is easily lost; in other words, these metals have low ionization energies.

While electronegativity increases along periods in the periodic table, and decreases down groups, electropositivity *decreases* along periods (from left to right) and *increases* down groups.

Intermolecular Force

Intermolecular forces (IMFs) are forces of attraction or repulsion which act between neighboring particles (atoms, molecules, or ions). They are weak compared to the intramolecular forces, the forces which keep a molecule together. For example the covalent bond, involving the sharing of electron pairs between atoms is much stronger than the forces present between the neighboring molecules. They are an essential part of force fields frequently used in molecular mechanics.

The investigation of intermolecular forces starts from macroscopic observations which point out the existence and action of forces at a molecular level. These observations include non-ideal-gas thermodynamic behavior reflected by virial coefficients, vapor pressure, viscosity, superficial tension and absorption data.

The first reference to the nature of microscopic forces is found in Alexis Clairaut's work *Theorie de la Figure de la Terre*. Other scientists who have contributed to the investigation of microscopic forces include: Laplace, Gauss, Maxwell and Boltzmann.

Attractive intermolecular forces are considered by the following types:

- Ion-induced dipole forces.
- Ion-dipole forces.
- Van der Waals forces (Keesom force, Debye force, and London dispersion force).

Information on intermolecular force is obtained by macroscopic measurements of properties like viscosity, PVT data. The link to microscopic aspects is given by virial coefficients and Lennard-Jones potentials.

Ion-dipole and Ion-induced Dipole Forces

Ion-dipole and ion-induced dipole forces are similar to dipole-dipole and induced-dipole interactions but involve ions, instead of only polar and non-polar molecules. Ion-dipole and ion-induced dipole forces are stronger than dipole-dipole interactions because the charge of any ion is much greater than the charge of a dipole moment. Ion-dipole bonding is stronger than hydrogen bonding.

An ion-dipole force consists of an ion and a polar molecule interacting. They align so that the positive and negative groups are next to one another, allowing for maximum attraction.

An ion-induced dipole force consists of an ion and a non-polar molecule interacting. Like a dipole-induced dipole force, the charge of the ion causes distortion of the electron cloud on the non-polar molecule.

Hydrogen Bonding

A hydrogen bond is the attraction between the lone pair of an electronegative atom and a hydrogen atom that is bonded to either nitrogen, oxygen, or fluorine. The hydrogen bond is often described as a strong electrostatic dipole-dipole interaction. However, it also has some features of covalent bonding: it is directional, stronger than a van der Waals interaction, produces interatomic distances shorter than the sum of van der Waals radius, and usually involves a limited number of interaction partners, which can be interpreted as a kind of valence.

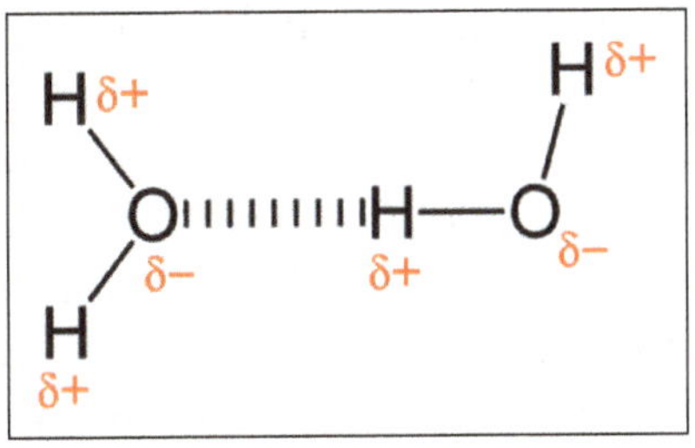

Intermolecular hydrogen bonding is responsible for the high boiling point of water (100 °C) compared to the other group 16 hydrides, which have no hydrogen bonds. Intramolecular hydrogen oxygen bonding is partly responsible for the secondary, tertiary, and quaternary structures of proteins and nucleic acids. It also plays an important role in the structure of polymers, both synthetic and natural.

Van Der Waals Forces

The van der Waals forces arise from interaction between uncharged atoms or molecules, leading not only to such phenomena as the cohesion of condensed phases and physical adsorption of gases, but also to a universal force of attraction between macroscopic bodies.

Keesom (Permanent-permanent Dipoles) Interaction

The first contribution to van der Waals forces is due to electrostatic interactions between charges (in molecular ions), dipoles (for polar molecules), quadrupoles (all molecules with symmetry lower than cubic), and permanent multipoles. It is referred to as Keesom interactions(named after Willem Hendrik Keesom). These forces originate from the attraction between permanent dipoles (dipolar molecules) and are temperature dependent.

They consist of attractive interactions between dipoles that are ensemble averaged over different rotational orientations of the dipoles. It is assumed that the molecules are constantly rotating and never get locked into place. This is a good assumption, but at some point molecules do get locked into place. The energy of a Keesom interaction depends on the inverse sixth power of the distance, unlike the interaction energy of two spatially fixed dipoles, which depends on the inverse third power of the distance. The Keesom interaction can only occur among molecules that possess permanent dipole moments a.k.a. two polar molecules. Also Keesom interactions are very weak van der Waals interactions and do not occur in aqueous solutions that contain electrolytes. The angle averaged interaction is given by the following equation:

$$\frac{-m_1^2 m_2^2}{24\pi^2 \varepsilon_o^2 \varepsilon_r^2 k_b T r^6} = V$$

Where m = charge per length, ε_o = permitivity of free space, ε_r = dielectric constant of surrounding material, T = temperature, k_b = Boltzmann constant, and r = distance between molecules.

Debye (Permanent-induced Dipoles) Force

The second contribution is the induction (also known as polarization) or Debye force, arising from interactions between rotating permanent dipoles and from the polarizability of atoms and molecules (induced dipoles). These induced dipoles occur when one molecule with a permanent dipole repels another molecule's electrons. A molecule with permanent dipole can induce a dipole in a similar neighboring molecule and cause mutual attraction. Debye forces cannot occur between atoms. The forces between induced and permanent dipoles are not as temperature dependent as Keesom interactions because the induced dipole is free to shift and rotate around the non-polar molecule. The Debye induction effects and Keesom orientation effects are referred to as polar interactions.

The induced dipole forces appear from the induction (also known as polarization), which is the attractive interaction between a permanent multipole on one molecule with an induced (by the former di/multi-pole) multipole on another. This interaction is called the Debye force, named after Peter J.W. Debye.

One example of an induction-interaction between permanent dipole and induced dipole is the interaction between HCl and Ar. In this system, Ar experiences a dipole as its electrons are attracted (to the H side of HCl) or repelled (from the Cl side) by HCl. The angle averaged interaction is given by the following equation.

$$\frac{-m_1^2 \alpha_2}{16\pi^2 \varepsilon_o^2 \varepsilon_r^2 r^6} = V$$

Where α = polarizability

This kind of interaction can be expected between any polar molecule and non-polar/symmetrical molecule. The induction-interaction force is far weaker than dipole-dipole interaction, but stronger than the London dispersion force.

London Dispersion Force (Dipole-induced Dipoles Interaction)

The third and dominant contribution is the dispersion or London force (fluctuating dipole-induced dipole), which arises due to the non-zero instantaneous dipole moments of all atoms and molecules. Such polarization can be induced either by a polar molecule or by the repulsion of negatively charged electron clouds in non-polar molecules. Thus, London interactions are caused by random fluctuations of electron density in an electron cloud. An atom with a large number of electrons will have a greater associated London force than an atom with fewer electrons. The dispersion (London) force is the most important component because all materials are polarizable, whereas Keesom and Debye forces require permanent dipoles. The London interaction is universal and is present in atom-atom interactions as well. For various reasons, London interactions (dispersion) have been considered relevant for interactions between macroscopic bodies in condensed systems. Hamaker developed the theory of van der Waals between macroscopic bodies in 1937 and showed that the additivity of these interactions renders them considerably more long-range.

Relative Strength of Forces

Bond type	Dissociation energy (kcal/mol)
Ionic Lattice Energy	250–4000
Covalent Bond Energy	30–260
Hydrogen Bonds	1–12 (about 5 in water)
Dipole–Dipole	0.5–2
London Dispersion Forces	<1 to 15 (estimated from the enthalpies of vaporization of hydrocarbons)

Note: this comparison is only approximate – the actual relative strengths will vary depending on the molecules involved. Ionic and covalent bonding will always be stronger than intermolecular forces in any given substance.

Effect on the Behavior of Gases

Intermolecular forces are repulsive at short distances and attractive at long distances. In a gas, the repulsive force chiefly has the effect of keeping two molecules from occupying the same volume. This gives a real gas a tendency to occupy a larger volume than an ideal gas at the same temperature and pressure. The attractive force draws molecules closer together and gives a real gas a tendency to occupy a smaller volume than an ideal gas. Which interaction is more important depends on temperature and pressure.

In a gas, the distances between molecules are generally large, so intermolecular forces have only a small effect. The attractive force is not overcome by the repulsive force, but by the thermal energy of the molecules. Temperature is the measure of thermal energy, so increasing temperature reduces the influence of the attractive force. In contrast, the influence of the repulsive force is essentially unaffected by temperature.

When a gas is compressed to increase its density, the influence of the attractive force increases. If the gas is made sufficiently dense, the attractions can become large enough to overcome the tendency of thermal motion to cause the molecules to spread out. Then the gas can condense to form a solid or liquid (i.e., a condensed phase). Lower temperature favors the formation of a condensed phase. In a condensed phase, there is very nearly a balance between the attractive and repulsive forces.

Quantum Mechanical Theories

Intermolecular forces observed between atoms and molecules can be described phenomenologically as occurring between permanent and instantaneous dipoles, as outlined above. Alternatively, one may seek a fundamental, unifying theory that is able to explain the various types of interactions such as hydrogen bonding, van der Waals forces and dipole-dipole interactions. Typically, this is done by applying the ideas of quantum mechanics to molecules, and Rayleigh–Schrödinger perturbation theory has been especially effective in this regard. When applied to existing quantum chemistry methods, such a quantum mechanical explanation of intermolecular interactions, this provides an array of approximate methods that can be used to analyze intermolecular interactions.

References

- SW Rick; SJ Stuart (2002). "Electronegativity equalization models". In Kenny B. Lipkowitz; Donald B. Boyd. Reviews in computational chemistry. Wiley. p. 106. ISBN 0-471-21576-7.
- Robert G. Parr; Weitao Yang (1994). Density-functional theory of atoms and molecules. Oxford University Press. p. 91. ISBN 0-19-509276-7.
- Atkins, Peter; Loretta Jones (1997). Chemistry: Molecules, Matter and Change. New York: W. H. Freeman & Co. pp. 294–295. ISBN 0-7167-3107-X.
- Campbell, Neil A.; Williamson, Brad; Heyden, Robin J. (2006). Biology: Exploring Life. Boston, MA: Pearson Prentice Hall. ISBN 0-13-250882-6. Retrieved 01, May 2020
- March, Jerry (1992). Advanced Organic Chemistry: Reactions, Mechanisms, and Structure. John Wiley & Sons. ISBN 0-471-60180-2.

5

Phase Transformation

Phase transformation takes place when one material changes its structure or composition. Phase diagram is used to represent the physical states of a substance under the influence of different conditions of temperature and pressure. All the diverse principles of phase transformation have been carefully analyzed in this chapter.

Phase

In the physical sciences, a phase is a region of space (a thermodynamic system), throughout which all physical properties of a material are essentially uniform. Examples of physical properties include density, index of refraction, magnetization and chemical composition. A simple description is that a phase is a region of material that is chemically uniform, physically distinct, and (often) mechanically separable. In a system consisting of ice and water in a glass jar, the ice cubes are one phase, the water is a second phase, and the humid air over the water is a third phase. The glass of the jar is another separate phase.

A small piece of rapidly melting argon ice shows the transition from solid to liquid.

The term *phase* is sometimes used as a synonym for state of matter, but there can be several immiscible phases of the same state of matter. Also, the term *phase* is sometimes used to refer to a set of equilibrium states demarcated in terms of state variables such as pressure and temperature by a phase boundary on a phase diagram. Because phase boundaries relate to changes in the organization of matter, such as a change from liquid to solid or a more subtle change from one crystal structure to another, this latter usage is similar to the use of "phase" as a synonym for state of matter. However, the state of matter and phase diagram usages are not commensurate with the formal definition given above and the intended meaning must be determined in part from the context in which the term is used.

Types of Phases

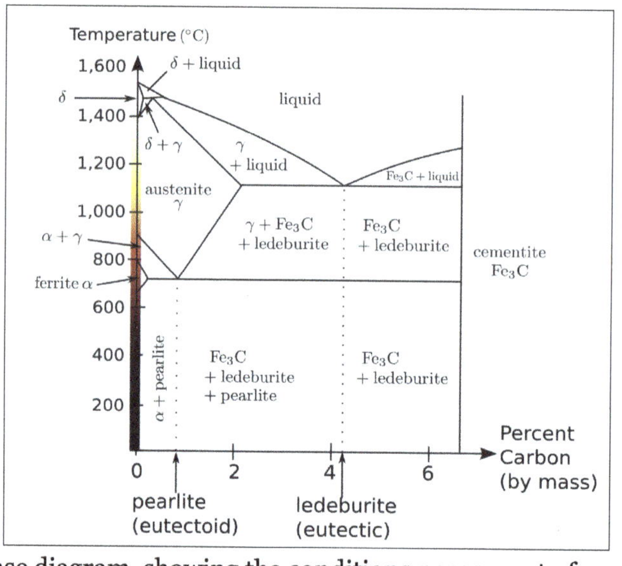

Iron-carbon phase diagram, showing the conditions necessary to form different phases.

Distinct phases may be described as different states of matter such as gas, liquid, solid, plasma or Bose–Einstein condensate. Useful mesophases between solid and liquid form other states of matter.

Distinct phases may also exist within a given state of matter. As shown in the diagram for iron alloys, several phases exist for both the solid and liquid states. Phases may also be differentiated based on solubility as in polar (hydrophilic) or non-polar (hydrophobic). A mixture of water (a polar liquid) and oil (a non-polar liquid) will spontaneously separate into two phases. Water has a very low solubility (is insoluble) in oil, and oil has a low solubility in water. Solubility is the maximum amount of a solute that can dissolve in a solvent before the solute ceases to dissolve and remains in a separate phase. A mixture can separate into more than two liquid phases and the concept of phase separation extends to solids, i.e., solids can form solid solutions or crystallize into distinct crystal phases. Metal pairs that are mutually soluble can form alloys, whereas metal pairs that are mutually insoluble cannot.

As many as eight immiscible liquid phases have been observed. Mutually immiscible liquid phases are formed from water (aqueous phase), hydrophobic organic solvents, perfluorocarbons (fluorous phase), silicones, several different metals, and also from molten phosphorus. Not all organic solvents are completely miscible, e.g. a mixture of ethylene glycol and toluene may separate into two distinct organic phases.

Phases do not need to macroscopically separate spontaneously. Emulsions and colloids are examples of immiscible phase pair combinations that do not physically separate.

Phase Equilibrium

Left to equilibration, many compositions will form a uniform single phase, but depending on the temperature and pressure even a single substance may separate into two or more distinct phases. Within each phase, the properties are uniform but between the two phases properties differ.

Water in a closed jar with an air space over it forms a two phase system. Most of the water is in the liquid phase, where it is held by the mutual attraction of water molecules. Even at equilibrium molecules are constantly in motion and, once in a while, a molecule in the liquid phase gains enough kinetic energy to break away from the liquid phase and enter the gas phase. Likewise, every once in a while a vapor molecule collides with the liquid surface and condenses into the liquid. At equilibrium, evaporation and condensation processes exactly balance and there is no net change in the volume of either phase.

At room temperature and pressure, the water jar reaches equilibrium when the air over the water has a humidity of about 3%. This percentage increases as the temperature goes up. At 100 °C and atmospheric pressure, equilibrium is not reached until the air is 100% water. If the liquid is heated a little over 100 °C, the transition from liquid to gas will occur not only at the surface, but throughout the liquid volume: the water boils.

Number of Phases

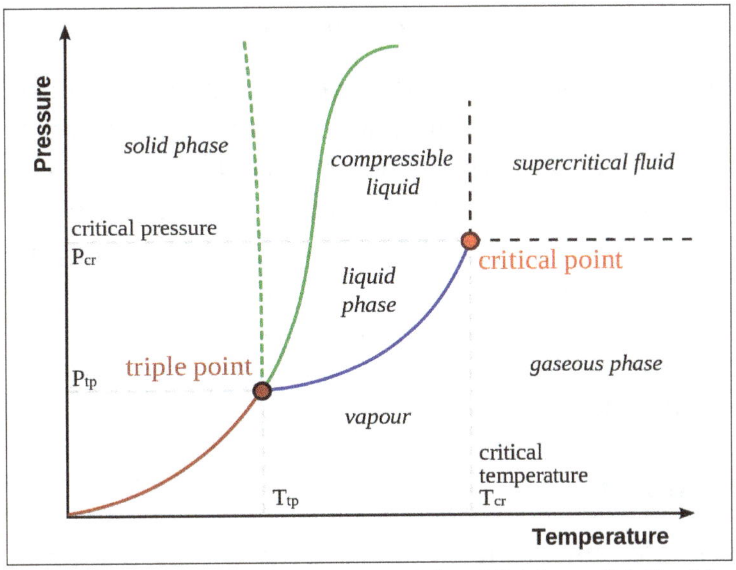

A typical phase diagram for a single-component material, exhibiting solid, liquid and gaseous phases. The solid green line shows the usual shape of the liquid–solid phase line. The dotted green line shows the anomalous behavior of water when the pressure increases. The triple point and the critical point are shown as red dots.

For a given composition, only certain phases are possible at a given temperature and pressure. The number and type of phases that will form is hard to predict and is usually determined by experiment. The results of such experiments can be plotted in phase diagrams.

The phase diagram shown here is for a single component system. In this simple system, which phases that are possible depends only on pressure and temperature. The markings show points where two or more phases can co-exist in equilibrium. At temperatures and pressures away from the markings, there will be only one phase at equilibrium.

In the diagram, the blue line marking the boundary between liquid and gas does not continue indefinitely, but terminates at a point called the critical point. As the temperature and pressure approach the critical point, the properties of the liquid and gas become progressively more similar. At the critical point, the liquid and gas become indistinguishable. Above the critical point, there are no longer separate liquid and gas phases: there is only a generic fluid phase referred to as a supercritical fluid. In water, the critical point occurs at around 647 K (374 °C or 705 °F) and 22.064 MPa.

An unusual feature of the water phase diagram is that the solid–liquid phase line (illustrated by the dotted green line) has a negative slope. For most substances, the slope is positive as exemplified by the dark green line. This unusual feature of water is related to ice having a lower density than liquid water. Increasing the pressure drives the water into the higher density phase, which causes melting.

Another interesting though not unusual feature of the phase diagram is the point where the solid–liquid phase line meets the liquid–gas phase line. The intersection is referred to as the triple point. At the triple point, all three phases can coexist.

Experimentally, the phase lines are relatively easy to map due to the interdependence of temperature and pressure that develops when multiple phases forms. Consider a test apparatus consisting of a closed and well insulated cylinder equipped with a piston. By charging the right amount of water and applying heat, the system can be brought to any point in the gas region of the phase diagram. If the piston is slowly lowered, the system will trace a curve of increasing temperature and pressure within the gas region of the phase diagram. At the point where gas begins to condense to liquid, the direction of the temperature and pressure curve will abruptly change to trace along the phase line until all of the water has condensed.

Interfacial Phenomena

Between two phases in equilibrium there is a narrow region where the properties are not that of either phase. Although this region may be very thin, it can have significant and easily observable effects, such as causing a liquid to exhibit surface tension. In mixtures, some components may preferentially move toward the interface. In terms of modeling, describing, or understanding the behavior of a particular system, it may be efficacious to treat the interfacial region as a separate phase.

Crystal Phases

A single material may have several distinct solid states capable of forming separate phases. Water is a well-known example of such a material. For example, water ice is ordinarily found in the hexagonal form ice I_h, but can also exist as the cubic ice I_c, the rhombohedral ice II, and many other forms. Polymorphism is the ability of a solid to exist in more than one crystal form. For pure chemical elements, polymorphism is known as allotropy. For example, diamond, graphite, and fullerenes are different allotropes of carbon.

Phase Transitions

When a substance undergoes a phase transition (changes from one state of matter to another) it usually either takes up or releases energy. For example, when water evaporates, the increase in kinetic energy as the evaporating molecules escape the attractive forces of the liquid is reflected in a decrease in temperature. The energy required to induce the phase transition is taken from the internal thermal energy of the water, which cools the liquid to a lower temperature; hence evaporation is useful for cooling. The reverse process, condensation, releases heat. The heat energy, or enthalpy, associated with a solid to liquid transition is the enthalpy of fusion and that associated with a solid to gas transition is the enthalpy of sublimation.

Phase Diagram

A phase diagram in physical chemistry, engineering, mineralogy, and materials science is a type of chart used to show conditions (pressure, temperature, volume, etc.) at which thermodynamically distinct phases occur and coexist at equilibrium.

Overview

Common components of a phase diagram are *lines of equilibrium* or *phase boundaries*, which refer to lines that mark conditions under which multiple phases can coexist at equilibrium. Phase transitions occur along lines of equilibrium.

Triple points are points on phase diagrams where lines of equilibrium intersect. Triple points mark conditions at which three different phases can coexist. For example, the water phase diagram has a triple point corresponding to the single temperature and pressure at which solid, liquid, and gaseous water can coexist in a stable equilibrium (273.16 K and a partial vapor pressure of 611.657 Pa).

The solidus is the temperature below which the substance is stable in the solid state. The liquidus is the temperature above which the substance is stable in a liquid state. There may be a gap between the solidus and liquidus; within the gap, the substance consists of a mixture of crystals and liquid (like a "slurry").

Types

2-dimensional Diagrams

The simplest phase diagrams are pressure–temperature diagrams of a single simple substance, such as water. The axes correspond to the pressure and temperature. The phase diagram shows, in pressure–temperature space, the lines of equilibrium or phase boundaries between the three phases of solid, liquid, and gas.

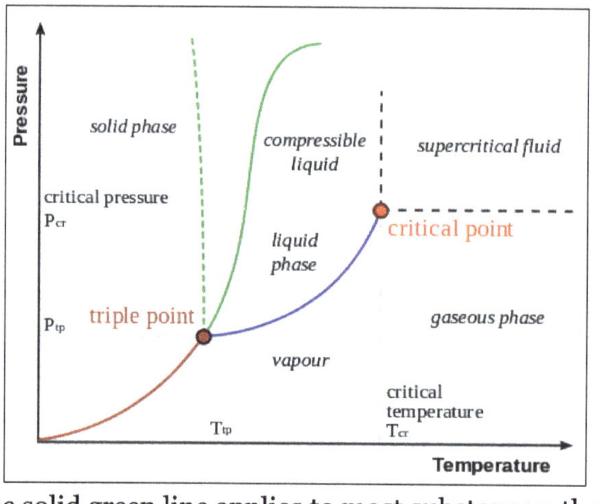

A typical phase diagram. The solid green line applies to most substances; the dotted green line gives the anomalous behavior of water. The green lines mark the freezing point and the blue line the boiling point, showing how they vary with pressure.

The fusion curves for water and Antimony are with negative slopes, All other liquids have positive slope on P-T diagram.

The curves on the phase diagram show the points where the free energy (and other derived properties) becomes non-analytic: their derivatives with respect to the coordinates (temperature and pressure in this example) change discontinuously (abruptly). For example, the heat capacity of a container filled with ice will change abruptly as the container is heated past the melting point. The open spaces, where the free energy is analytic, correspond to single phase regions. Single phase regions are separated by lines of non-analytical behavior, where phase transitions occur, which are called phase boundaries.

In the diagram on the left, the phase boundary between liquid and gas does not continue indefinitely. Instead, it terminates at a point on the phase diagram called the critical point. This reflects the fact that, at extremely high temperatures and pressures, the liquid and gaseous phases become indistinguishable, in what is known as a supercritical fluid. In water, the critical point occurs at around T_c = 647.096 K (373.946 °C), p_c = 22.064 MPa (217.75 atm) and ρ_c = 356 kg/m³.

The existence of the liquid–gas critical point reveals a slight ambiguity in labelling the single phase regions. When going from the liquid to the gaseous phase, one usually crosses the phase boundary, but it is possible to choose a path that never crosses the boundary by going to the right of the critical point. Thus, the liquid and gaseous phases can blend continuously into each other. The solid–liquid phase boundary can only end in a critical point if the solid and liquid phases have the same symmetry group.

The solid–liquid phase boundary in the phase diagram of most substances has a positive slope; the greater the pressure on a given substance, the closer together the molecules of the substance are brought to each other, which increases the effect of the substance's intermolecular forces. Thus, the substance requires a higher temperature for its molecules to have enough energy to break out of the fixed pattern of the solid phase and enter the liquid phase. A similar concept applies to liquid–gas phase changes. Water, because of its particular properties, is one of the several exceptions to the rule.

Other Thermodynamic Properties

In addition to temperature and pressure, other thermodynamic properties may be graphed in phase diagrams. Examples of such thermodynamic properties include specific volume, specific enthalpy, or specific entropy. For example, single-component graphs of temperature vs. specific entropy (*T* vs. *s*) for water/steam or for a refrigerant are commonly used to illustrate thermodynamic cycles such as a Carnot cycle, Rankine cycle, or vapor-compression refrigeration cycle.

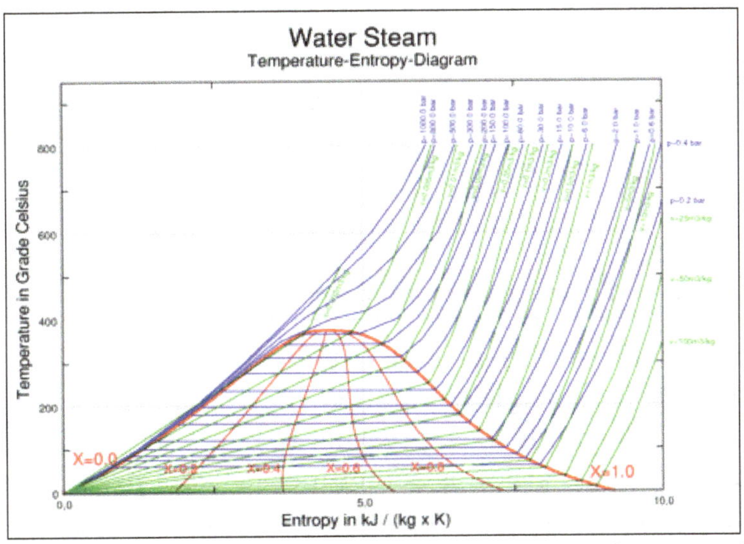

Temperature vs. specific entropy phase diagram for water/steam. In the area under the red dome, liquid water and steam coexist in equilibrium. The critical point is at the top of the dome. Liquid water is to the left of the dome. Steam is to the right of the dome. The blue lines/curves are isobars showing constant pressure. The green lines/curves are isochors showing constant specific volume. The red curves show constant quality.

In a two-dimensional graph, two of the thermodynamic quantities may be shown on the horizontal and vertical axes. Additional thermodynamic quantities may each be illustrated in increments as a series of lines - curved, straight, or a combination of curved and straight. Each of these iso-lines represents the thermodynamic quantity at a certain constant value.

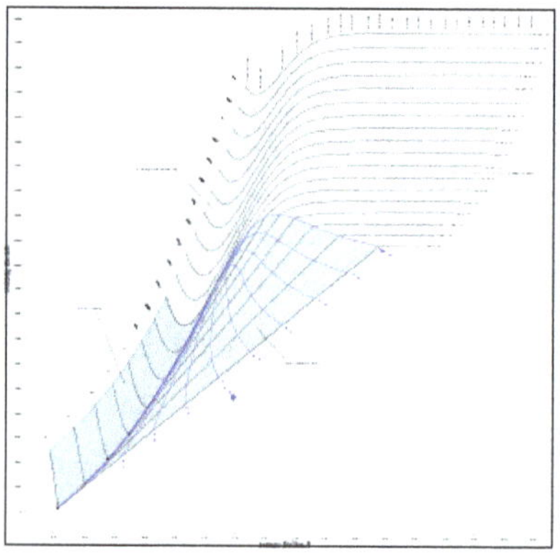

enthalpy–entropy (h–s) diagram for steam.

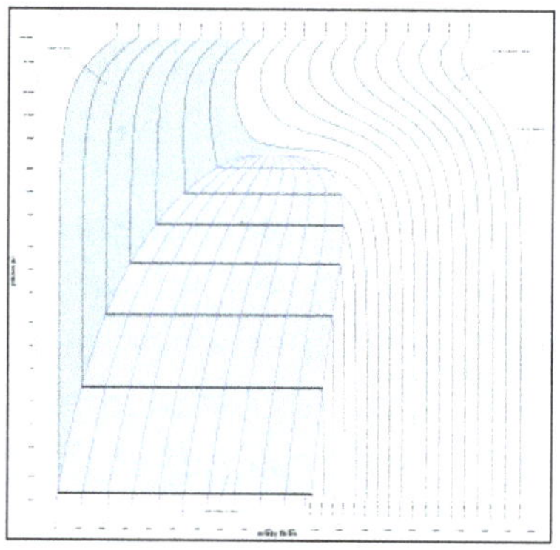

pressure–enthalpy (p–h) diagram for steam.

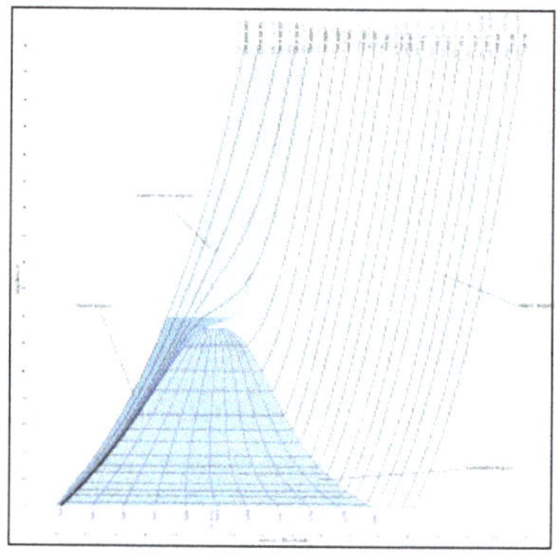

temperature–entropy (T–s) diagram for steam.

3-dimensional Diagrams

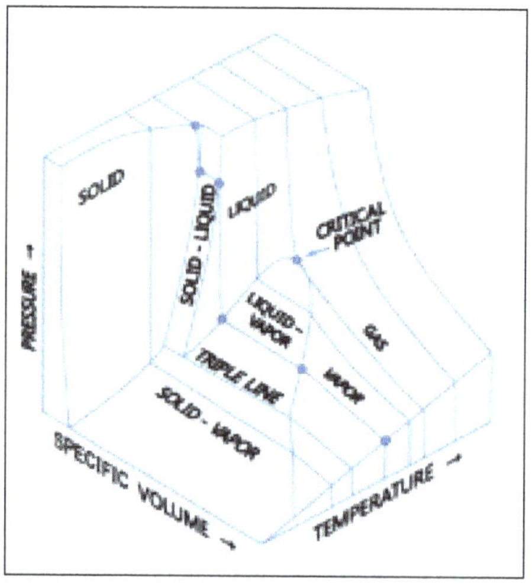

p–v–T 3D diagram for fixed amount of pure material.

It is possible to envision three-dimensional (3D) graphs showing three thermodynamic quantities. For example, for a single component, a 3D Cartesian coordinate type graph can show temperature (T) on one axis, pressure (p) on a second axis, and specific volume (v) on a third. Such a 3D graph is sometimes called a p–v–T diagram. The equilibrium conditions are shown as curves on a curved surface in 3D with areas for solid, liquid, and vapor phases and areas where solid and liquid, solid and vapor, or liquid and vapor coexist in equilibrium. A line on the surface called a triple line is where solid, liquid and vapor can all coexist in equilibrium. The critical point remains a point on the surface even on a 3D phase diagram.

For water, the 3D p–v–T diagram is seen here:

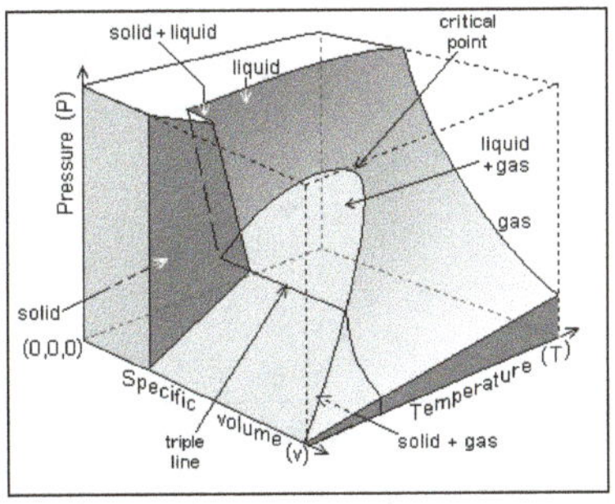

3D phase diagram of water fluids and selected ices.

An orthographic projection of the 3D p–v–T graph showing pressure and temperature as the vertical and horizontal axes collapses the 3D plot into the standard 2D pressure–temperature diagram. When this is done, the solid–vapor, solid–liquid, and liquid–vapor surfaces collapse into three corresponding curved lines meeting at the triple point, which is the collapsed orthographic projection of the triple line.

Binary Mixtures

Other much more complex types of phase diagrams can be constructed, particularly when more than one pure component is present. In that case, concentration becomes an important variable. Phase diagrams with more than two dimensions can be constructed that show the effect of more than two variables on the phase of a substance. Phase diagrams can use other variables in addition to or in place of temperature, pressure and composition, for example the strength of an applied electrical or magnetic field, and they can also involve substances that take on more than just three states of matter.

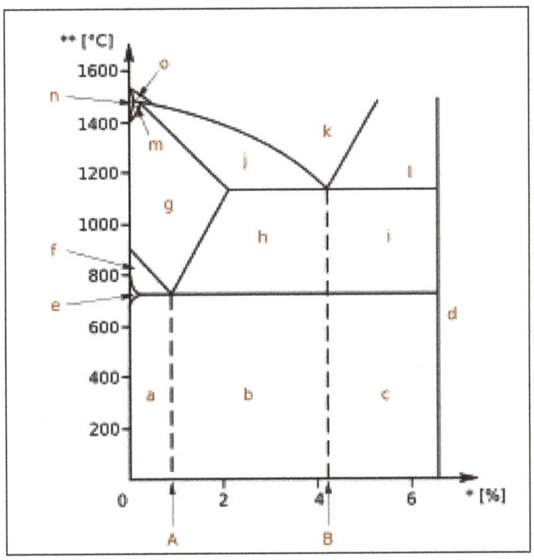

The iron–iron carbide (Fe–Fe_3C) phase diagram. The percentage of carbon present and the temperature define the phase of the iron carbon alloy and therefore its physical characteristics and mechanical properties. The percentage of carbon determines the type of the ferrous alloy: iron, steel or cast iron.

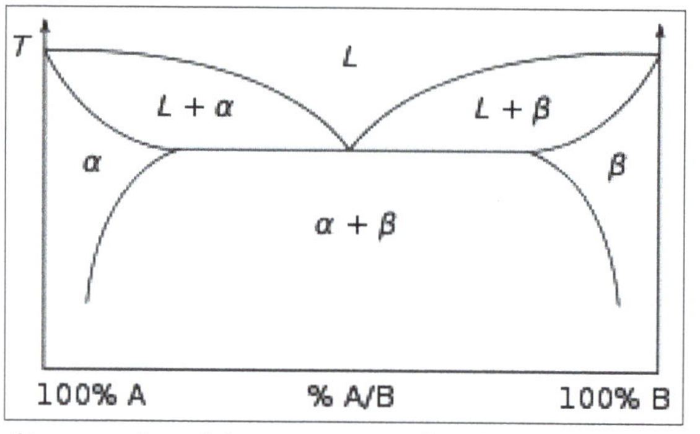

A phase diagram for a binary system displaying a eutectic point.

One type of phase diagram plots temperature against the relative concentrations of two substances in a binary mixture called a *binary phase diagram*, as shown at right. Such a mixture can be either a solid solution, eutectic or peritectic, among others. These two types of mixtures result in very different graphs. Another type of binary phase diagram is a *boiling-point diagram* for a mixture of two components, i. e. chemical compounds. For two particular volatile components at a certain pressure such as atmospheric pressure, a boiling-point diagram shows what vapor (gas) compositions are in equilibrium with given liquid compositions depending on temperature. In a typical binary boiling-point diagram, temperature is plotted on a vertical axis and mixture composition on a horizontal axis.

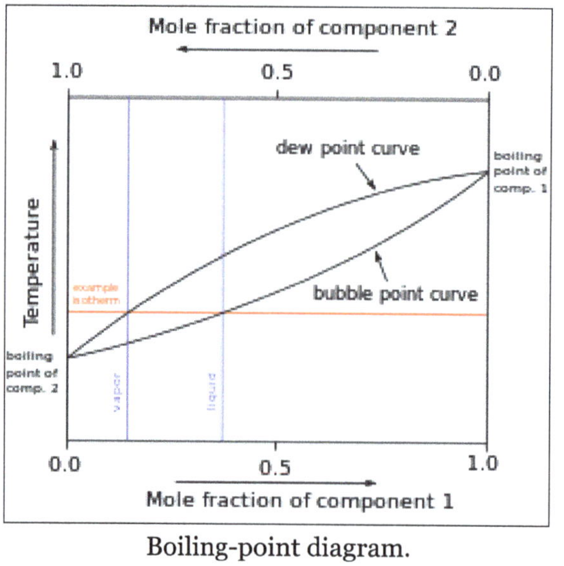

Boiling-point diagram.

A simple example diagram with hypothetical components 1 and 2 in a non-azeotropic mixture is shown at right. The fact that there are two separate curved lines joining the boiling points of the pure components means that the vapor composition is usually not the same as the liquid composition the vapor is in equilibrium with.

In addition to the above-mentioned types of phase diagrams, there are thousands of other possible combinations. Some of the major features of phase diagrams include congruent points, where a solid phase transforms directly into a liquid. There is also the peritectoid, a point where two solid phases combine into one solid phase during cooling. The inverse of this, when one solid phase transforms into two solid phases during cooling, is called the eutectoid.

A complex phase diagram of great technological importance is that of the iron–carbon system for less than 7% carbon.

The x-axis of such a diagram represents the concentration variable of the mixture. As the mixtures are typically far from dilute and their density as a function of temperature is usually unknown, the preferred concentration measure is mole fraction. A volume-based measure like molarity would be inadvisable.

Crystals

Polymorphic and polyamorphic substances have multiple crystal or amorphous phases, which can be graphed in a similar fashion to solid, liquid, and gas phases.

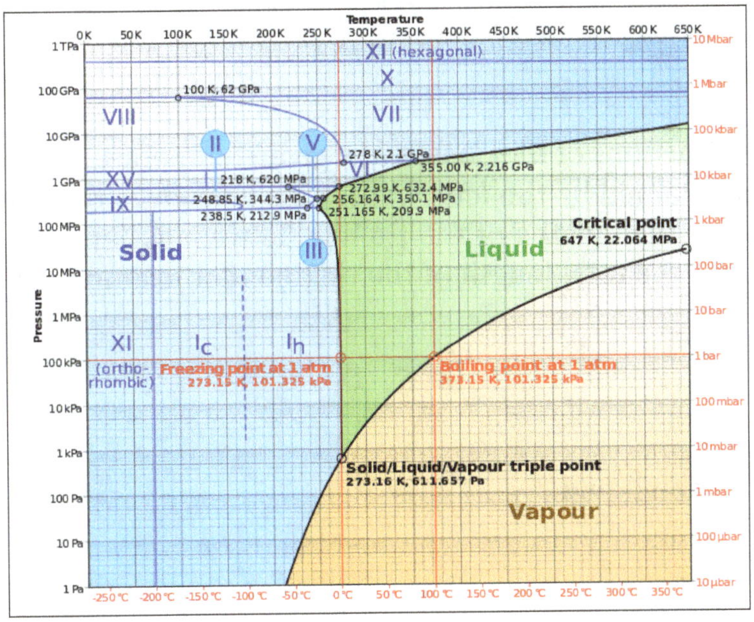

Log-lin pressure–temperature phase diagram of water. The Roman numerals indicate various ice phases. A similar diagram may be found on the site Water structure and science.

Mesophases

Some organic materials pass through intermediate states between solid and liquid; these states are called mesophases. Attention has been directed to mesophases because they enable display devices and have become commercially important through the so-called liquid-crystal technology. Phase diagrams are used to describe the occurrence of mesophases.

- The composition of phases in the two-phase region is not same.

- To find the composition of the individual phases in the two- phase region, a horizontal line (XY), called *tie line*, is drawn and its intercepts on the liquidus and solidus lines, C_l and C_s, are taken as the composition of the liquid and solid respectively.

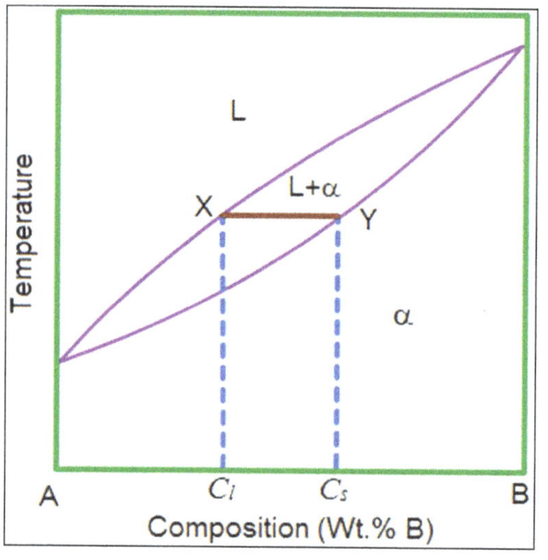

Lever Rule

- The relative fractions of the phases at a given temperature for an alloy composition C_O is obtained by the *lever rule*. This rule gives the fraction of a phase by the ratio of the lengths of the tie line between C_O and composition of the other phase to the total length of the tie line. For example, fraction solid, f_S is given by,

$$f_s = \frac{MX}{XY} = \frac{C_o - C_1}{C_s - C_l}$$

Similarly fraction liquid, f_l

$$f_l = \frac{MY}{XY} = \frac{C_s - C_o}{C_s - C_l}$$

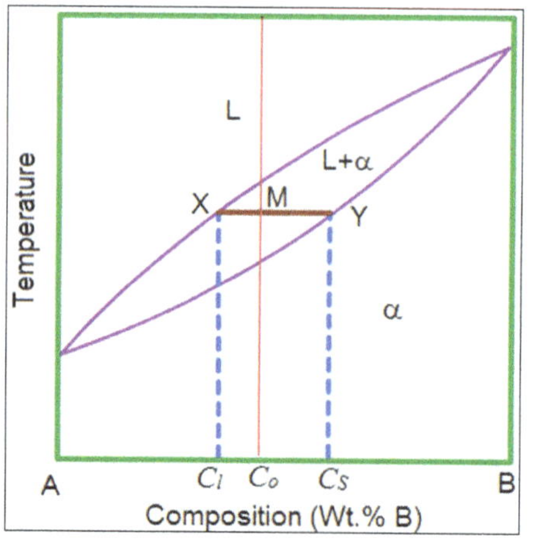

Cooling Curves

- Upon cooling from liquid state, the temperature of the pure metal (A or B) drops continuously till melting point at which solidification starts. Solidification happens at a constant temperature (line PQ) as F =0 (F = 1 – 2 +1 = 0). The temperature drops again on completion of solidification.
- For any alloy (1, 2, 3 etc.) temp. drops till the liquidus (L_1, L_2, L_3). However, in this case, solidification proceeds over a range of temperature as F = 1 (2 – 2 + 1 = 1). Once solidification completes at the solidus (S_1, S_2, S_3) the temp. drops again.

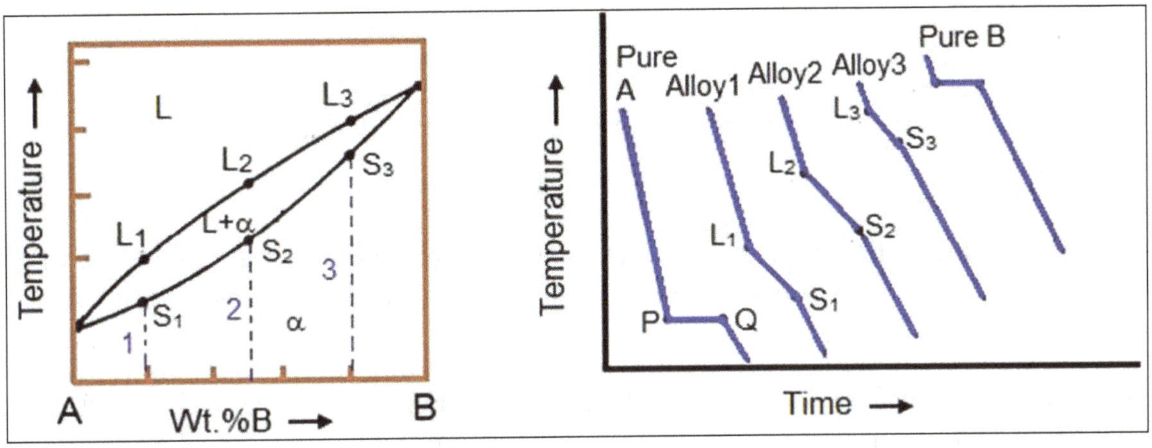

Phase Diagrams- Limited Solubility

- Not all metals are completely soluble in each other. Distinctions can be made between two types solid solutions with limited solubility – (i) Eutectic and (ii) Peritectic.
- When the melting points of two metals are comparable, a eutectic system forms while a peritectic results when melting points are significantly different.
- A eutectic reaction is defined as the one which generates two solids from the liquid at a given temperature and composition, L→ α + β.
- Peritectic is Liquid + Solid 1 → Solid 2 (L+ α → β).
- In both the cases three phases (two solids and a liquid) coexist and the degrees of freedom F = 2 – 3 + 1 = 0. This is known as invariant (F = 0) reaction or transformation.

Eutectic Phase diagram

- In the eutectic system between two metals A and B, two solid solutions, one rich in A (α) and another rich in B (β) form.

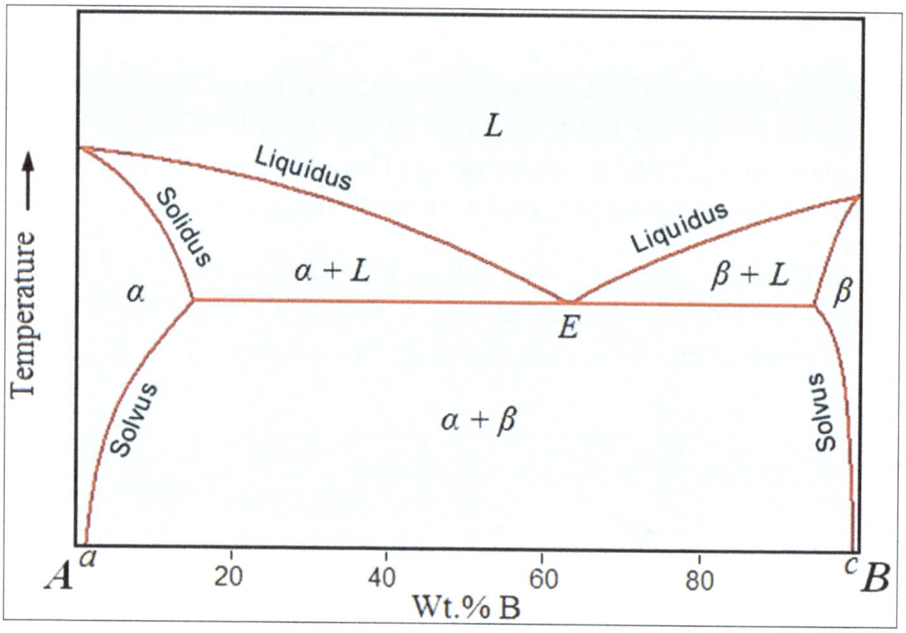

- In addition to liquidus and solidus lines there are two more lines on A and B rich ends which define the solubility limits B in A and A in B respectively. These are called solvus lines.

- Three phases ($L+\alpha+\beta$) coexist at point *E*. This point is called eutectic point or composition. Left of *E* is called hypoeutectic whereas right of E is called hypereutectic.

- A eutectic composition solidifies as a eutectic mixture of α and β phases. The microstructure at room temperature (RT) may consist of alternate layers or lamellae of α and β.

- In hypoeutectic alloys the α phase solidifies first and the microstructure at RT consists of this α phase (called proeutectic α) and the eutectic ($\alpha+\beta$) mixture. Similarly hypereutectic alloys consist of proeutectic β and the eutectic mixture.

- The melting point at the eutectic point is minimum. That's why Pb-Sn eutectic alloys are used as solders. Other eutectic systems are Ag-Cu, Al-Si, Al-Cu.

Eutectic Cooling Curves

- While cooling a hypoeutectic alloy from the liquid state, the temp. drops continuously till liquidus point, *a*, at which crystals of proeutectic α begins to form.

- On further cooling the fraction of α increases. At any point, *b*, in the two-phase region the α fraction is given by the *lever rule* as *bn/mn*.

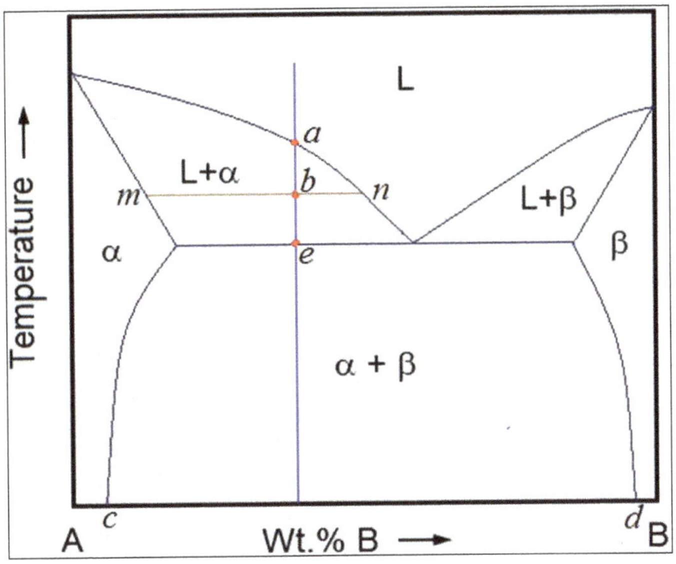

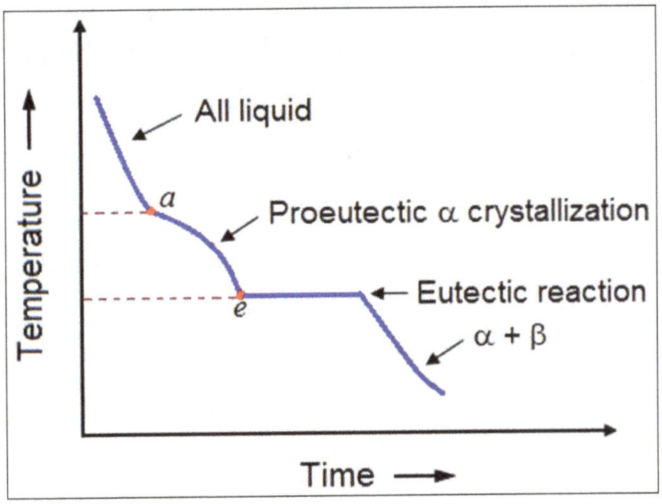

- Solidification of proeutectic α continues till the eutectic temperature is reached. The inflection in the cooling curve between points *a* and *e* is due to evolution of the latent heat.
- At the eutectic point (*e*) the solidification of eutectic mixture ($\alpha + \beta$) begins through the eutectic reaction and proceeds at a constant temperature as F = 0 (2 – 3 + 1).
- The cooling behavior in hypereutectic alloy is similar except that proeutectic β forms below the liquidus.
- For a eutectic composition, the proeutectic portion is absent and the cooling curve appears like that of a pure metal.
- Any composition left of point *c* or right of point *d* (α and β single phase region respectively) will cool and solidify like an isomorphous system.

Peritectic Phase Diagram

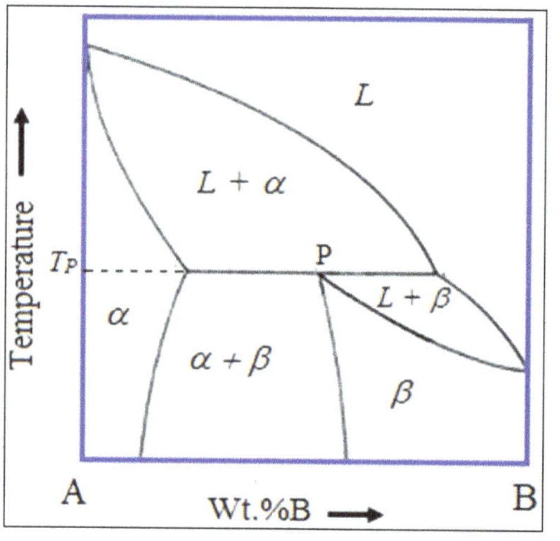

- $L + \alpha \rightarrow \beta$. An alloy cooling slowly through the peritectic point, *P*, the α phase will crystallize first just below the liquidus line. At the peritectic temperature, T_P all of the liquid and α will convert to β .
- Any composition left of *P* will generate excess α and similarly compositions right of *P* will give rise to an excess of liquid.
- Peritectic systems – Pt - Ag, Ni - Re, Fe - Ge, Sn-Sb (babbit).

Monotectic Phase Diagram

- Another three phase invariant reaction that occurs in some binary system is monotectic reaction in which a liquid transforms to another liquid and a solid. $L_1 \rightarrow L_2 + \alpha$.
- Two liquids are immiscible like water and oil over certain range of compositions. Cu-Pb system has a monotectic at 36% Pb and 955 °C.

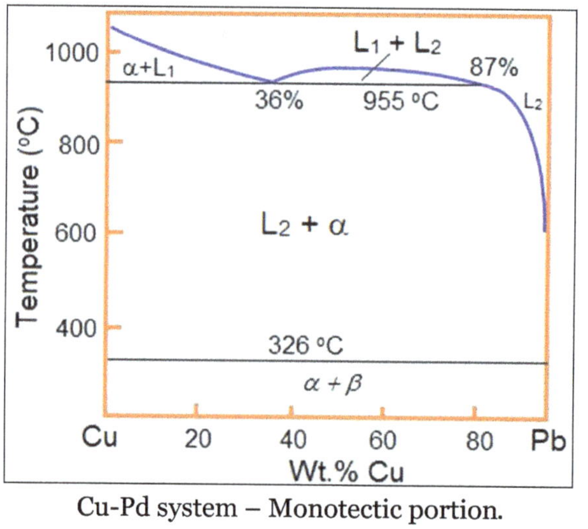

Cu-Pd system – Monotectic portion.

Phase Diagrams with Intermediate Phases

- Binary system can have two types of solid solutions/phases terminal phases and intermediate phases.
- Terminal phases occur near the pure metal ends, e.g. α and β phases in the eutectic system.
- Intermediate phases occur inside the phase diagram and are separated by two-phase regions.
- The Cu-Zn system contains both types of phases. α and η are terminal phases and β, γ, δ and ε are intermediate phases.
- Intermediate phases form in ceramic phase diagrams also. For example, in the $Al_2O_3 - SiO_2$ system an intermediate phase called mullite ($3Al_2O_3 - 2SiO_2$) is formed.

Intermediate Phases - Cu-Zn Phase Diagram

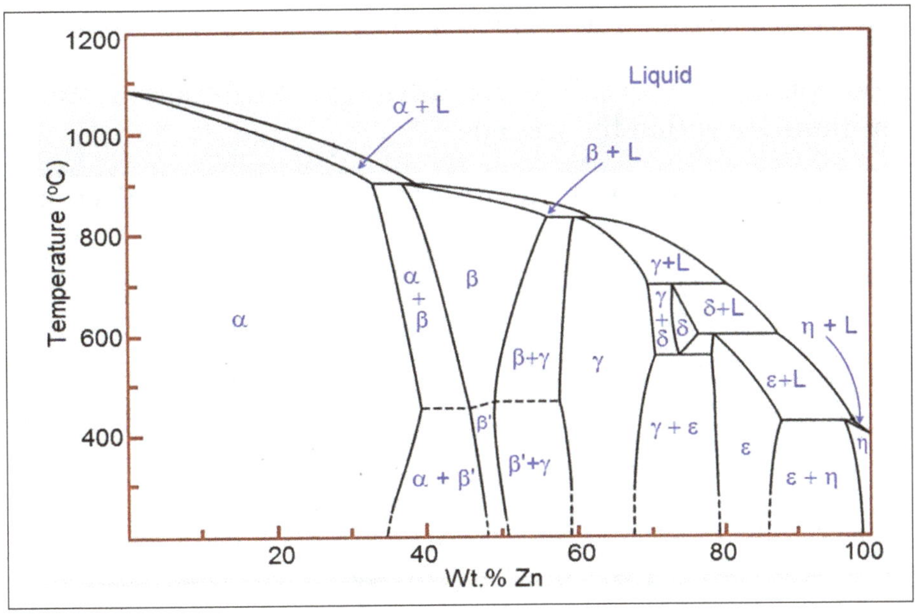

Cu-Zn phase diagram. α and η are terminal phases and β, γ, δ and ε are intermediate phases.

Phase Diagrams with Compounds

- Sometimes a crystalline compound called *intermetallic compound* may form between two metals.
- Such compounds generally have a distinct chemical formula or stoichiometry.
- Example – Mg_2Pb in the Mg-Pb system (appear as a vertical line at 81% Pb), Mg_2Ni, Mg_2Si, Fe_3C.

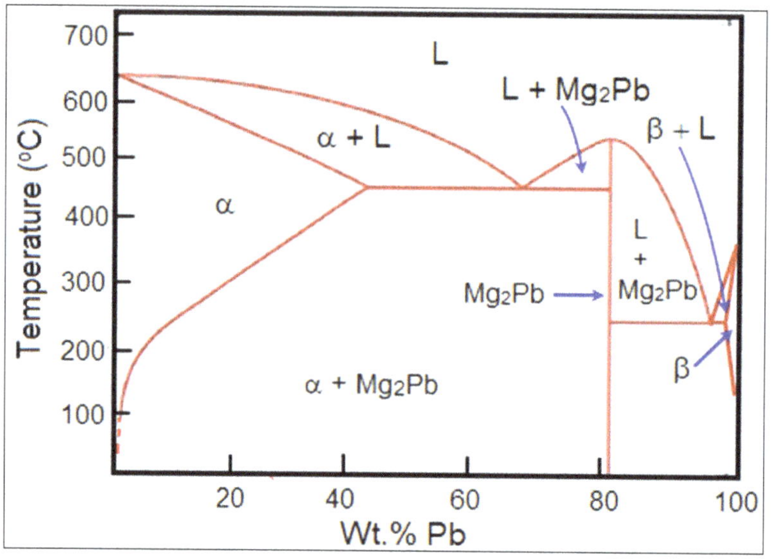

Ternary Phase Diagram

- A ternary or three component phase diagram has the form of an triangular prism with an equilateral triangle as a base.
- Pure components are at each vertex, sides are binary compositions and ternary compositions are within the triangle.
- The composition lines on the triangle is constructed from projections of surfaces.

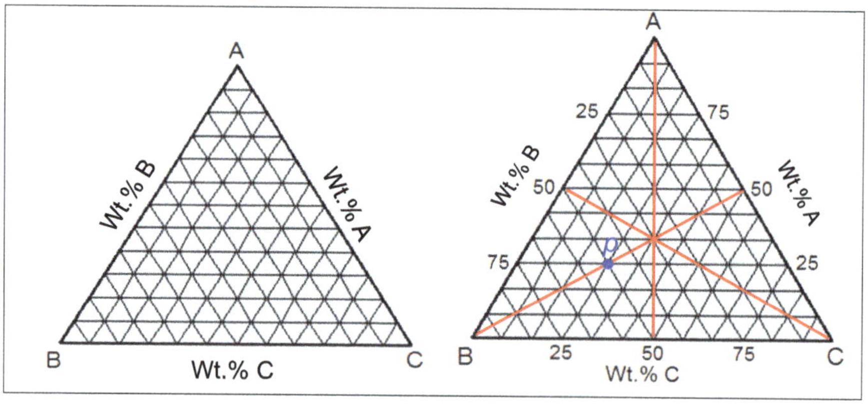

- The temperature varies along the height of the prism. The composition triangle is an isothermal section. Alternatively projections of different surfaces and lines can be shown as temperature contours.
- The composition of any point in the triangle is determined by drawing perpendiculars from corners to the opposite sides and measuring the distance of the point along the perpendicular.
- Point *p*, for example, lies on the isocomoposition line 25% A along the perpen-

dicular A-50. Hence, percentage of A in the alloy is 25%. Similarly B is 50% and C is 25%.

Phase Boundary

In thermal equilibrium, each phase (i.e. liquid, solid etc.) of physical matter comes to an end at a transitional point, or spatial interface, called a phase boundary, due to the immiscibility of said matter with the matter on the other side of said boundary. This immiscibility is due to at least one difference between the two substances' corresponding physical properties. The behavior of phase boundaries has been a developing subject of interest and an active research field, called interface science, in physics and mathematics for almost two centuries, due partly to phase boundaries naturally arising in many physical processes, such as the capillarity effect, the growth of grain boundaries, the physics of binary alloys, and the formation of snow flakes.

One of the oldest problems in the area dates back to Lamé and Clapeyron who studied the freezing of the ground. Their goal was to determine the thickness of solid crust generated by the cooling of a liquid at constant temperature filling the half-space. In 1889, Stefan, while working on the freezing of the ground developed these ideas further and formulated the two phase model which came to be known as the Stefan Problem.

The proof of existence and uniqueness of a solution to the Stefan problem was done in many stages. Proving the general existence and uniqueness of the solutions in the case of $d = 3$ was solved by Shoshana Kamin.

Phase Transition

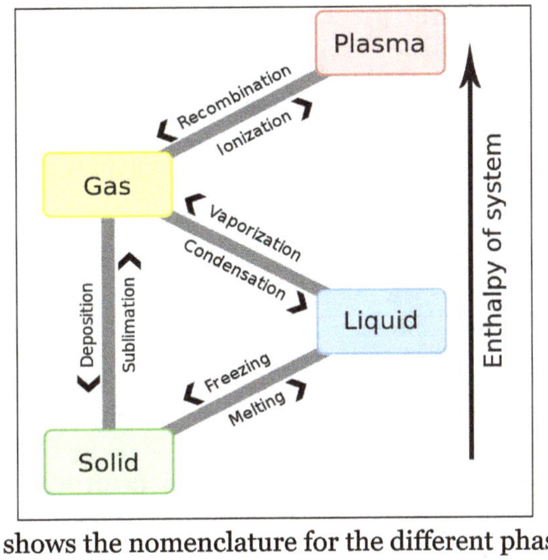

This diagram shows the nomenclature for the different phase transitions.

The term phase transition is most commonly used to describe transitions between solid, liquid and gaseous states of matter, and, in rare cases, plasma. A phase of a thermodynamic system and the states of matter have uniform physical properties. During a phase transition of a given medium certain properties of the medium change, often discontinuously, as a result of the change of some external condition, such as temperature, pressure, or others. For example, a liquid may become gas upon heating to the boiling point, resulting in an abrupt change in volume. The measurement of the external conditions at which the transformation occurs is termed the phase transition. Phase transitions are common in nature and used today in many technologies.

Types of Phase Transition

Examples of phase transitions include:

- The transitions between the solid, liquid, and gaseous phases of a single component, due to the effects of temperature and/or pressure:

		To			
		Solid	**Liquid**	**Gas**	**Plasma**
From	**Solid**	Solid-solid transformation	Melting	Sublimation	—
	Liquid	Freezing	—	Boiling/evaporation	—
	Gas	Deposition	Condensation	—	Ionization
	Plasma	—	—	Recombination/ deionization	—

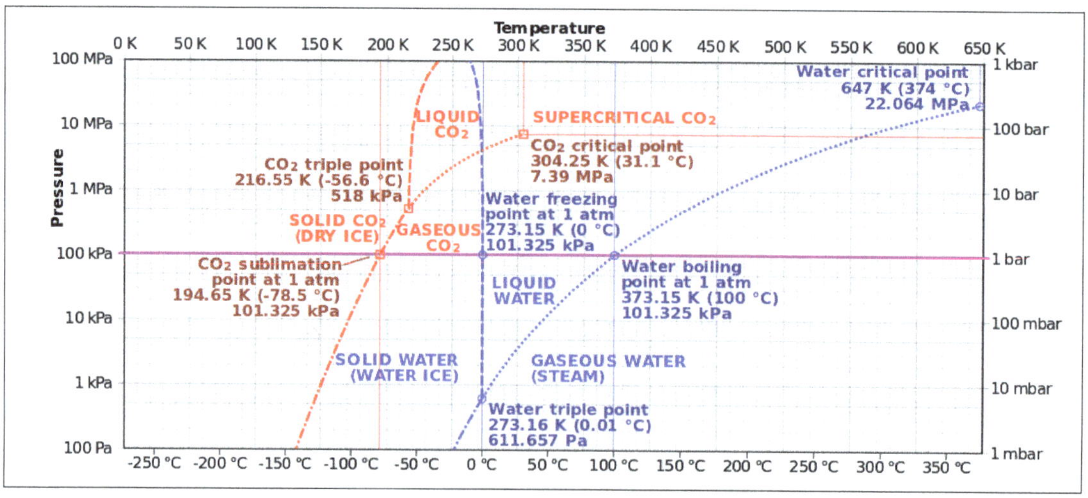

Comparison of phase diagrams of carbon dioxide (red) and water (blue) explaining their different phase transitions at 1 atmosphere.

- A eutectic transformation, in which a two component single phase liquid is cooled and transforms into two solid phases. The same process, but beginning with a solid instead of a liquid is called a eutectoid transformation.

- A peritectic transformation, in which a two component single phase solid is heated and transforms into a solid phase and a liquid phase.
- A spinodal decomposition, in which a single phase is cooled and separates into two different compositions of that same phase.
- Transition to a mesophase between solid and liquid, such as one of the "liquid crystal" phases.
- The transition between the ferromagnetic and paramagnetic phases of magnetic materials at the Curie point.
- The transition between differently ordered, commensurate or incommensurate, magnetic structures, such as in cerium antimonide.
- The martensitic transformation which occurs as one of the many phase transformations in carbon steel and stands as a model for displacive phase transformations.
- Changes in the crystallographic structure such as between ferrite and austenite of iron.
- Order-disorder transitions such as in alpha-titanium aluminides.
- The dependence of the adsorption geometry on coverage and temperature, such as for hydrogen on iron (110).
- The emergence of superconductivity in certain metals and ceramics when cooled below a critical temperature.
- The transition between different molecular structures (polymorphs, allotropes or polyamorphs), especially of solids, such as between an amorphous structure and a crystal structure, between two different crystal structures, or between two amorphous structures.
- Quantum condensation of bosonic fluids (Bose–Einstein condensation). The superfluid transition in liquid helium is an example of this.
- The breaking of symmetries in the laws of physics during the early history of the universe as its temperature cooled.
- Isotope fractionation occurs during a phase transition, the ratio of light to heavy isotopes in the involved molecules changes. When water vapor condenses (an equilibrium fractionation), the heavier water isotopes (18O and 2H) become enriched in the liquid phase while the lighter isotopes (16O and 1H) tend toward the vapor phase.

Phase transitions occur when the thermodynamic free energy of a system is non-analytic for some choice of thermodynamic variables (cf. phases). This condition generally stems from the interactions of a large number of particles in a system, and does not appear in systems that are too small. It is important to note that phase transitions can occur and are defined for non-thermodynamic systems, where temperature is not a parameter. Examples include: quantum phase transitions, dynamic phase transitions, and topological (structural) phase transitions. In these types of systems other parameters take the place of temperature. For instance, connection probability replaces temperature for percolating networks.

At the phase transition point (for instance, boiling point) the two phases of a substance, liquid and vapor, have identical free energies and therefore are equally likely to exist. Below the boiling point, the liquid is the more stable state of the two, whereas above the gaseous form is preferred.

It is sometimes possible to change the state of a system diabatically (as opposed to adiabatically) in such a way that it can be brought past a phase transition point without undergoing a phase transition. The resulting state is metastable, i.e., less stable than the phase to which the transition would have occurred, but not unstable either. This occurs in superheating, supercooling, and supersaturation, for example.

Classifications

Ehrenfest Classification

Paul Ehrenfest classified phase transitions based on the behavior of the thermodynamic free energy as a function of other thermodynamic variables. Under this scheme, phase transitions were labeled by the lowest derivative of the free energy that is discontinuous at the transition. *First-order phase transitions* exhibit a discontinuity in the first derivative of the free energy with respect to some thermodynamic variable. The various solid/liquid/gas transitions are classified as first-order transitions because they involve a discontinuous change in density, which is the (inverse of the) first derivative of the free energy with respect to pressure. *Second-order phase transitions* are continuous in the first derivative (the order parameter, which is the first derivative of the free energy with respect to the external field, is continuous across the transition) but exhibit discontinuity in a second derivative of the free energy. These include the ferromagnetic phase transition in materials such as iron, where the magnetization, which is the first derivative of the free energy with respect to the applied magnetic field strength, increases continuously from zero as the temperature is lowered below the Curie temperature. The magnetic susceptibility, the second derivative of the free energy with the field, changes discontinuously. Under the Ehrenfest classification scheme, there could in principle be third, fourth, and higher-order phase transitions.

Though useful, Ehrenfest's classification has been found to be an incomplete method of classifying phase transitions, for it does not take into account the case where a derivative of free energy diverges (which is only possible in the thermodynamic limit).

For instance, in the ferromagnetic transition, the heat capacity diverges to infinity.The same phenomena is also seen in superconducting phase transition.

Modern Classifications

In the modern classification scheme, phase transitions are divided into two broad categories, named similarly to the Ehrenfest classes:

First-order phase transitions are those that involve a latent heat. During such a transition, a system either absorbs or releases a fixed (and typically large) amount of energy per volume. During this process, the temperature of the system will stay constant as heat is added: the system is in a "mixed-phase regime" in which some parts of the system have completed the transition and others have not. Familiar examples are the melting of ice or the boiling of water (the water does not instantly turn into vapor, but forms a turbulent mixture of liquid water and vapor bubbles). Imry and Wortis showed that quenched disorder can broaden a first-order transition in that the transformation is completed over a finite range of temperatures, but phenomena like supercooling and superheating survive and hysteresis is observed on thermal cycling.

Second-order phase transitions are also called *continuous phase transitions*. They are characterized by a divergent susceptibility, an infinite correlation length, and a power-law decay of correlations near criticality. Examples of second-order phase transitions are the ferromagnetic transition, superconducting transition (for a Type-I superconductor the phase transition is second-order at zero external field and for a Type-II superconductor the phase transition is second-order for both normal state-mixed state and mixed state-superconducting state transitions) and the superfluid transition. In contrast to viscosity, thermal expansion and heat capacity of amorphous materials show a relatively sudden change at the glass transition temperature which enables accurate detection using differential scanning calorimetry measurements. Lev Landau gave a phenomenological theory of second-order phase transitions.

Apart from isolated, simple phase transitions, there exist transition lines as well as multicritical points, when varying external parameters like the magnetic field or composition.

Several transitions are known as the *infinite-order phase transitions*. They are continuous but break no symmetries. The most famous example is the Kosterlitz–Thouless transition in the two-dimensional XY model. Many quantum phase transitions, e.g., in two-dimensional electron gases, belong to this class.

The liquid–glass transition is observed in many polymers and other liquids that can be supercooled far below the melting point of the crystalline phase. This is atypical in several respects. It is not a transition between thermodynamic ground states: it is widely believed that the true ground state is always crystalline. Glass is a *quenched disorder* state, and its entropy, density, and so on, depend on the thermal history. Therefore, the glass transition is primarily a dynamic phenomenon: on cooling a liquid, internal degrees of freedom successively fall out of equilibrium. Some theoretical methods predict an underlying phase transition in the hypothetical limit of infinitely long relaxation times. No direct experimental evidence supports the existence of these transitions.

Characteristic Properties

Phase Coexistence

A disorder-broadened first-order transition occurs over a finite range of temperatures where the fraction of the low-temperature equilibrium phase grows from zero to one (100%) as the temperature is lowered. This continuous variation of the coexisting fractions with temperature raised interesting possibilities. On cooling, some liquids vitrify into a glass rather than transform to the equilibrium crystal phase. This happens if the cooling rate is faster than a critical cooling rate, and is attributed to the molecular motions becoming so slow that the molecules cannot rearrange into the crystal positions. This slowing down happens below a glass-formation temperature Tg, which may depend on the applied pressure., If the first-order freezing transition occurs over a range of temperatures, and Tg falls within this range, then there is an interesting possibility that the transition is arrested when it is partial and incomplete. Extending these ideas to first-order magnetic transitions being arrested at low temperatures, resulted in the observation of incomplete magnetic transitions, with two magnetic phases coexisting, down to the lowest temperature. First reported in the case of a ferromagnetic to anti-ferromagnetic transition, such persistent phase coexistence has now been reported across a variety of first-order magnetic transitions. These include colossal-magnetoresistance manganite materials, magnetocaloric materials, magnetic shape memory materials, and other materials. The interesting feature of these observations of Tg falling within the temperature range over which the transition occurs is that the first-order magnetic transition is influenced by magnetic field, just like the structural transition is influenced by pressure. The relative ease with which magnetic field can be controlled, in contrast to pressure, raises the possibility that one can study the interplay between Tg and Tc in an exhaustive way. Phase coexistence across first-order magnetic transitions will then enable the resolution of outstanding issues in understanding glasses.

Critical Points

In any system containing liquid and gaseous phases, there exists a special combination of pressure and temperature, known as the critical point, at which the transition between liquid and gas becomes a second-order transition. Near the critical point, the fluid is sufficiently hot and compressed that the distinction between the liquid and gaseous phases is almost non-existent. This is associated with the phenomenon of critical opalescence, a milky appearance of the liquid due to density fluctuations at all possible wavelengths (including those of visible light).

Symmetry

Phase transitions often involve a symmetry breaking process. For instance, the cooling of a fluid into a crystalline solid breaks continuous translation symmetry: each point in the fluid has the same properties, but each point in a crystal does not have the same properties (unless the points are chosen from the lattice points of the crystal lattice). Typically, the high-temperature phase contains more symmetries than the low-temperature phase due to spontaneous symmetry breaking, with the exception of certain accidental symmetries (e.g. the formation of heavy virtual particles, which only occurs at low temperatures).

Order Parameters

An order parameter is a measure of the degree of order across the boundaries in a phase transition system; it normally ranges between zero in one phase (usually above the critical point) and nonzero in the other. At the critical point, the order parameter susceptibility will usually diverge.

An example of an order parameter is the net magnetization in a ferromagnetic system undergoing a phase transition. For liquid/gas transitions, the order parameter is the difference of the densities.

From a theoretical perspective, order parameters arise from symmetry breaking. When this happens, one needs to introduce one or more extra variables to describe the state of the system. For example, in the ferromagnetic phase, one must provide the net magnetization, whose direction was spontaneously chosen when the system cooled below the Curie point. However, note that order parameters can also be defined for non-symmetry-breaking transitions. Some phase transitions, such as superconducting and ferromagnetic, can have order parameters for more than one degree of freedom. In such phases, the order parameter may take the form of a complex number, a vector, or even a tensor, the magnitude of which goes to zero at the phase transition.

There also exist dual descriptions of phase transitions in terms of disorder parameters. These indicate the presence of line-like excitations such as vortex- or defect lines.

Relevance in Cosmology

Symmetry-breaking phase transitions play an important role in cosmology. It has been speculated by Lee Smolin and Benjamin and Jeremy Bernstein that, in the hot early universe, the vacuum (i.e. the various quantum fields that fill space) possessed a large number of symmetries. As the universe expanded and cooled, the vacuum underwent a series of symmetry-breaking phase transitions. For example, the electroweak transition broke the SU(2)×U(1) symmetry of the electroweak field into the U(1) symmetry of the present-day electromagnetic field. This transition is important to understanding the asymmetry between the amount of matter and antimatter in the present-day universe.

Progressive phase transitions in an expanding universe are implicated in the development of order in the universe, as is illustrated by the work of Eric Chaisson and David Layzer.

Critical Exponents and Universality Classes

Continuous phase transitions are easier to study than first-order transitions due to the absence of latent heat, and they have been discovered to have many interesting properties. The phenomena associated with continuous phase transitions are called critical phenomena, due to their association with critical points.

It turns out that continuous phase transitions can be characterized by parameters known as critical exponents. The most important one is perhaps the exponent describing the divergence of the thermal correlation length by approaching the transition. For instance, let us examine the behavior of the heat capacity near such a transition. We vary the temperature T of the system while keeping all the other thermodynamic variables fixed, and find that the transition occurs at some critical temperature T_c. When T is near T_c, the heat capacity C typically has a power law behavior,

$$C \propto |T_c - T|^{-\alpha}.$$

The heat capacity of amorphous materials has such a behaviour near the glass transition temperature where the universal critical exponent α = 0.59 A similar behavior, but with the exponent ν instead of α, applies for the correlation length.

The exponent ν is positive. This is different with α. Its actual value depends on the type of phase transition we are considering.

It is widely believed that the critical exponents are the same above and below the critical temperature. It has now been shown that this is not necessarily true: When a continuous symmetry is explicitly broken down to a discrete symmetry by irrelevant (in the renormalization group sense) anisotropies, then some exponents (such as γ, the exponent of the susceptibility) are not identical.

For −1 < α < 0, the heat capacity has a "kink" at the transition temperature. This is the behavior of liquid helium at the lambda transition from a normal state to the superfluid state, for which experiments have found α = -0.013±0.003. At least one experiment was performed in the zero-gravity conditions of an orbiting satellite to minimize pressure differences in the sample. This experimental value of α agrees with theoretical predictions based on variational perturbation theory.

For 0 < α < 1, the heat capacity diverges at the transition temperature (though, since α < 1, the enthalpy stays finite). An example of such behavior is the 3D ferromagnetic phase transition. In the three-dimensional Ising model for uniaxial magnets, detailed theoretical studies have yielded the exponent $\alpha \sim +0.110$.

Some model systems do not obey a power-law behavior. For example, mean field theory predicts a finite discontinuity of the heat capacity at the transition temperature, and the two-dimensional Ising model has a logarithmic divergence. However, these systems are limiting cases and an exception to the rule. Real phase transitions exhibit power-law behavior.

Several other critical exponents, β, γ, δ, ν, and η, are defined, examining the power law behavior of a measurable physical quantity near the phase transition. Exponents are related by scaling relations, such as,

$$\beta = \gamma / (\delta - 1), \qquad \nu = \gamma / (2 - \eta).$$

It can be shown that there are only two independent exponents, e.g. ν and η.

It is a remarkable fact that phase transitions arising in different systems often possess the same set of critical exponents. This phenomenon is known as *universality*. For example, the critical exponents at the liquid–gas critical point have been found to be independent of the chemical composition of the fluid.

More impressively, but understandably from above, they are an exact match for the critical exponents of the ferromagnetic phase transition in uniaxial magnets. Such systems are said to be in the same universality class. Universality is a prediction of the renormalization group theory of phase transitions, which states that the thermodynamic properties of a system near a phase transition depend only on a small number of features, such as dimensionality and symmetry, and are insensitive to the underlying microscopic properties of the system. Again, the divergence of the correlation length is the essential point.

Critical Slowing Down and other Phenomena

There are also other critical phenomena; e.g., besides *static functions* there is also *critical dynamics*. As a consequence, at a phase transition one may observe critical slowing down or *speeding up*. The large *static universality classes* of a continuous phase transition split into smaller *dynamic universality* classes. In addition to the critical exponents, there are also universal relations for certain static or dynamic functions of the magnetic fields and temperature differences from the critical value.

Percolation Theory

Another phenomenon which shows phase transitions and critical exponents is percolation. The simplest example is perhaps percolation in a two dimensional square lattice.

Sites are randomly occupied with probability p. For small values of p the occupied sites form only small clusters. At a certain threshold p_c a giant cluster is formed and we have a second-order phase transition. The behavior of P_∞ near p_c is, $P_\infty \sim (p-p_c)^\beta$, where β is a critical exponent.

Phase Transitions in Biological Systems

Phase transitions play many important roles in biological systems. Examples include the lipid bilayer formation, the coil-globule transition in the process of protein folding and DNA melting, liquid crystal-like transitions in the process of DNA condensation, and cooperative ligand binding to DNA and proteins with the character of phase transition.

In *biological membranes*, gel to liquid crystalline phase transitions play a critical role in physiological functioning of biomembranes. In gel phase, due to low fluidity of membrane lipid fatty-acyl chains, membrane proteins have restricted movement and thus are restrained in exercise of their physiological role. Plants depend critically on photosynthesis by chloroplast thylakoid membranes which are exposed cold environmental temperatures. Thylakoid membranes retain innate fluidity even at relatively low temperatures because of high degree of fatty-acyl disorder allowed by their high content of linolenic acid, 18-carbon chain with 3-double bonds. Gel-to-liquid crystalline phase transition temperature of biological membranes can be determined by many techniques including calorimetry, flouorescence, spin label electron paramagnetic resonance and NMR by recording measurements of the concerned parameter by at series of sample temperatures. A simple method for its determination from 13-C NMR line intensities has also been proposed.

It has been proposed that some biological systems might lie near critical points. Examples include neural networks in the salamander retina, bird flocks gene expression networks in Drosophila, and protein folding. However, it is not clear whether or not alternative reasons could explain some of the phenomena supporting arguments for criticality. It has also been suggested that biological organisms share two key properties of phase transitions: the change of macroscopic behavior and the coherence of a system at a critical point.

Homogeneous Nucleation

- In homogeneous nucleation the probability of nucleation is same throughout the volume of the parent phase.
- The simplest example of nucleation is solidification of a metal.
- Above melting point T_m, liquid free energy, $G_l < G_s$ (Solid free energy) and free energy change for solidification $\Delta G > 0$.

Below T_m, $\Delta G < 0$ and nuclei of the solid phase form.

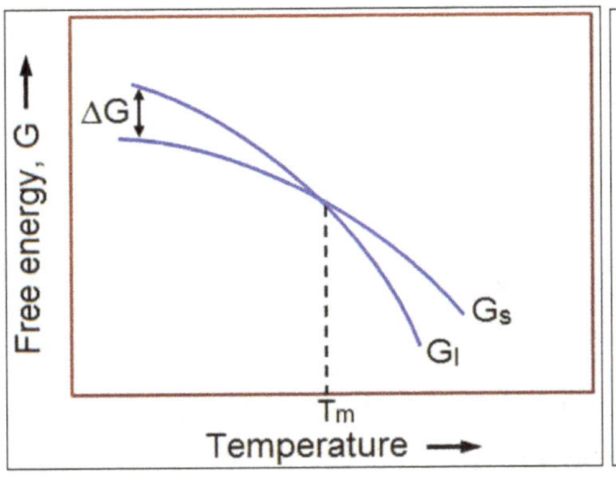

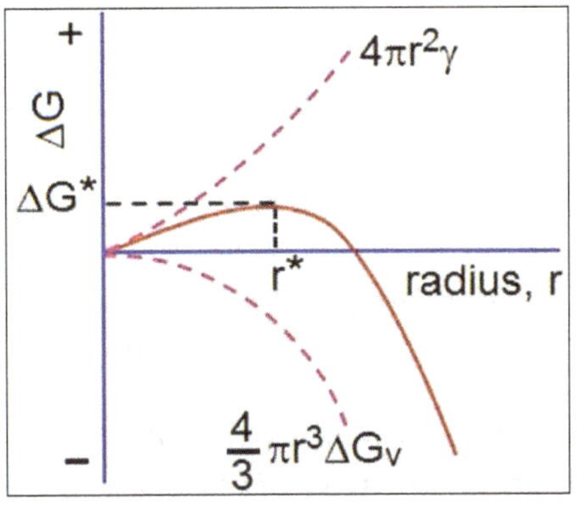

- There are two contributions to free energy change, volume free energy ΔG_v and surface free energy, γ due to creation of a new surface.
- Taking the nucleus as a spherical particle of radius r,

$$\Delta G = 4/3\pi r^2 \Delta G_v + 4\pi r^2 \gamma$$

- The tiny particle of the solid that forms first will be stable only when it achieves a critical radius (r*). Below the critical radius it is unstable and is called *embryo*.
- Since this happens at the maximum of the ΔG vs. r curve $d\Delta G / dr = 4\pi r\ \Delta G\ + 8\pi r\gamma = 0$
- This yields $r^* = -\dfrac{2\gamma}{\Delta G_v}$ and $\Delta G^* = \dfrac{16\pi\gamma^3}{3(\Delta G_v)^2}$
- Here, the probability of nucleation is much higher at certain preferred sites such as mold wall, inclusions, grain boundaries, compared to rest of the parent phase.
- Example - Solidification of a liquid on an inclusion surface,

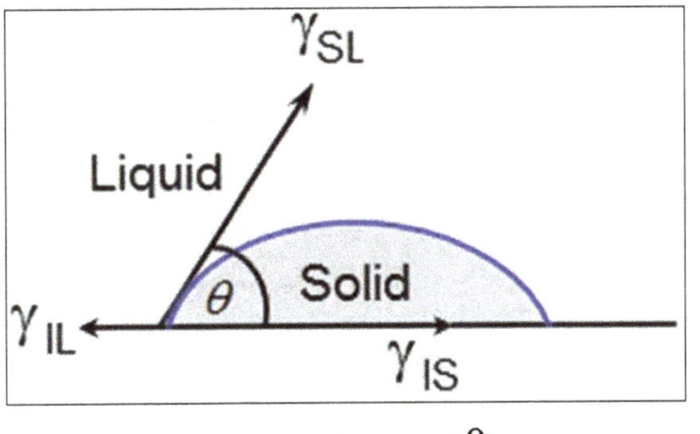

$$\gamma_{IL} = \gamma_{IS} + \gamma_{SL}\cos\theta$$

- With a similar approach it can be shown that $r^* = -\dfrac{2\gamma_{SL}}{\Delta G_v}$ and

$$\Delta G^*_{Het} = \frac{4\pi\gamma^3}{3(\Delta G_v)^2}(2 - 3\cos\theta + \cos^3\theta) = \Delta G^*_{Hom} S(\theta)$$

- The small value of θ ensures that the energy barrier (ΔG) is effectively lowered in heterogeneous nucleation.

Nucleation and Growth Kinetics

- Once the embryo exceeds the critical size r*, the growth of the nucleus starts. Nucleation continues simultaneously.
- Nucleation and growth rates are function of temp. Nucleation rate increases with cooling rate and degree of undercooling ($\Delta T = T_m - T$).
- High nucleation rate and low growth – Finer grain size.
- The over all transformation rate is the product of nucleation and growth rates.

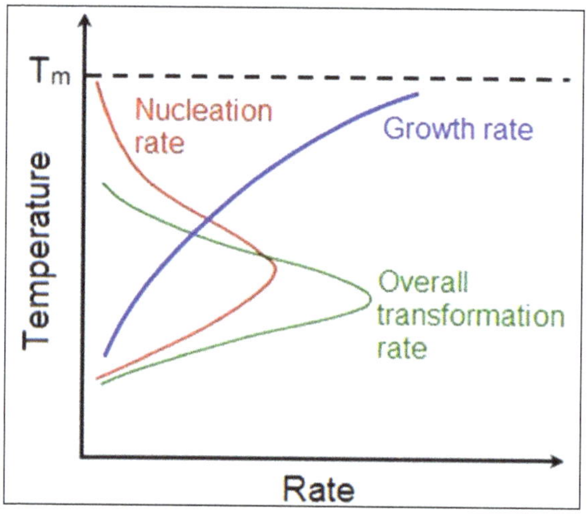

Fe-C Phase Diagram

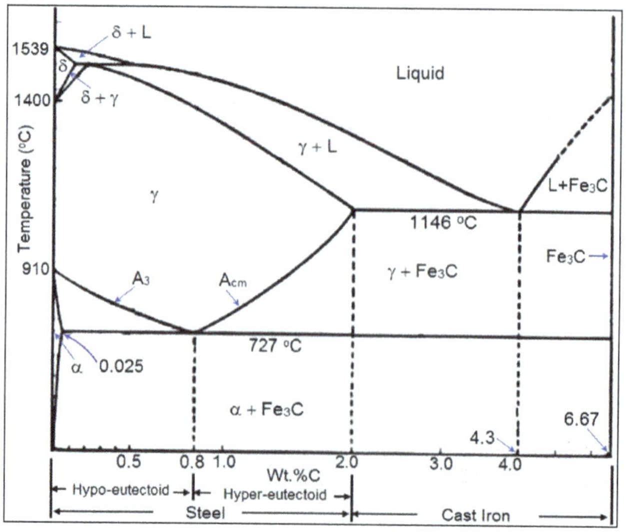

Phases in Fe-C System

- *α-ferrite* – Interstitial solid solution of C in BCC iron. Max solubility of C is 0.025%. Exists from 273°C to 910°C.
- *Austenite* (γ) - Interstitial solid solution of C in FCC iron. Max solubility of C is 2.1%. Exists from 910°C - 1394°C.
- *δ-ferrite* (BCC) exists over the temp range of 1394°C to 1539°C. Max solubility of C is 0.09%.
- Cementite, Fe_3C - is an intermetallic compound. C content in Fe_3C is 6.67%.
- Graphite, the free form of C, also exists in the Fe-C system.
- Bainite (*B*) is another phase which forms in steels at higher cooling rates.
- The hard phase martensite (*M*) forms below the bainitic temperature range at high cooling rates.

Critical Temperatures in Fe-C System

- The eutectoid temperature (727°C) during heating and cooling is A_{c1} and A_{r1} respectively. *A* for arrêt (arrest), *c* for chauffage (heating) and *r* for refroidissement (cooling).
- At normal rates of heating or cooling $A_{c1} > A_{r1}$.
- A_2 at 768 °C is the currie temp above which Fe turns paramagnetic while heating.
- The temperatures corresponding to $(\gamma + \alpha)/\gamma$ and $(\gamma + Fe_3C)/\gamma$ phase boundaries are function of carbon content and are represented as A_3 and A_{cm} respectively.
- The eutectic temperature is 1146 °C.
- The peritectic temperature is at 1495 °C.

Phase Transformation in Fe-C System

- Peritectic reaction at 1495 °C:

  L (0.53% C) + δ (0.09% C) → γ (0.17% C)

- Eutectic reaction at 1146°C:

  L (4.3% C) → γ (2.1 % C) + Fe_3C (6.67% C). The eutectic mixture of austenite (γ) and cementite (Fe_3C) is called *Ledeburite*. Compositions right and left of 4.3% are called hyper and hypoeutectic steels (Cast iron) respectively.

- Eutectoid reaction at 727°C:

  γ (0.8 % C) → α (0.025% C) + Fe_3C (6.67% C). The eutectoid mixture of ferrite (α) and cementite (Fe_3C) is called *Pearlite*.

- Compositions right and left of 0.8% are called hyper and hypoeutectoid steels respectively.

- Compositions up to 2.1% C are steels and beyond this it is considered as cast iron.

Microstructures

- A eutectoid steel (0.8% C) will have 100% pearlite (*p*) at room temperature (RT). The pearlite formed under equilibrium conditions consists of alternate lamellas of ferrite and Fe_3C.

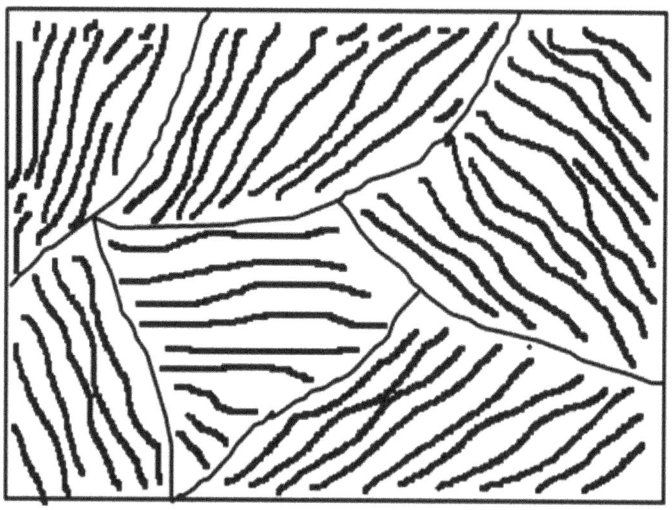

Schematic of Pearlite – White areas are α.

- Hypoeutectoid steels – $\alpha + p$; hypereutectoid – $Fe_3C + p$.

- Hypoeutectic cast irons consist of γ + *ledeburite* (*Le*) below the eutectic temp and $p + Fe_3C + Le$ at RT as the γ transforms to Fe_3C and p at the eutectoid temp. Similarly hyper eutectic cast irons will have a structure of $Fe_3C + Le$.

T-T-T Diagram

- The relation between temperature and time for the formation of a phase is given by T-T-T or temp – time – transformation diagrams also known as isothermal-transformation diagram.

- A typical T-T-T diagram is shown below. The phases formed on isothermal holding at a given temp for a certain period of time are indicated.

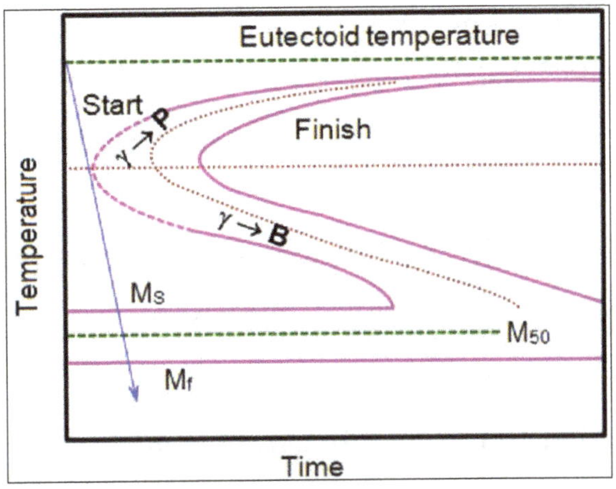

T-T-T diagram.

- At normal cooling rates pearlite (*P*) forms, higher cooling rates generates bainite (*B*). The size of pearlite or bainite depends on the transformation temp.
- Martensite (*M*) forms when the steel is cooled below the maternsite start (M_S) temp at much higher cooling rate so that the nose of the T-T-T curve (shown dotted) is avoided (the long blue arrow).
- Diffusion rates below M_S is so low that $\gamma \rightarrow M$ transformation is a diffusionless process (the C content remains same). However, the crystal structure changes from FCC (γ) to body centered tetragonal (BCT).

C-C-T Diagram

- In actual practice a steel is generally cooled continuously. Continuous-cooling-transformation (C-C-T) diagrams depict this situation.
- The C-C-T curve (Blue) is shifted to the right of the T-T-T (dashed) curve as continuous cooling transformation occurs at lower temperature and longer time compared isothermal holding.

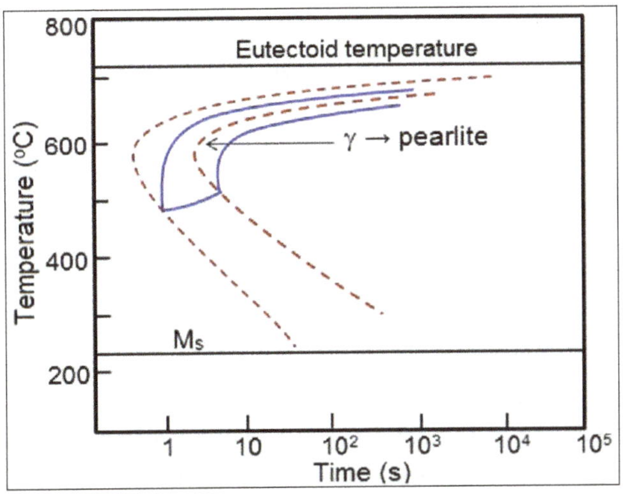

- Bainite generally does not form in steels during continuous cooling and hence the C-C-T curve ceases just below the nose.
- The microstructure (fine or coarse) depends on the cooling rate. Higher the cooling rate finer the microstructure is.
- Finer size pearlite is called *sorbite* and very fine size pearlite is called *troostite.*
- The critical cooling rate is the one at which the cooling curve just touches the nose of the C-C-T curve.
- A cooling rate higher than the critical rate is needed to form martensite.

Diffusionless Transformation

A diffusionless transformation is a phase change that occurs without the long-range diffusion of atoms but rather by some form of cooperative, homogeneous movement of many atoms that results in a change in crystal structure. These movements are small, usually less than the interatomic distances, and the atoms maintain their relative relationships. The ordered movement of large numbers of atoms lead some to refer to these as *military* transformations in contrast to *civilian* diffusion-based phase changes.

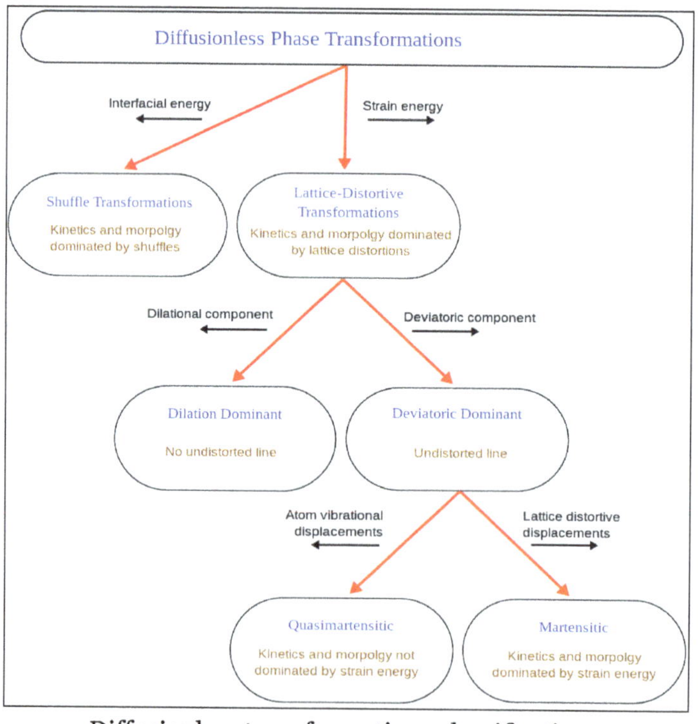

Diffusionless transformations classifications.

The most commonly encountered transformation of this type is the martensitic transformation which, while being the least studied, is only one subset of non-diffusional transformations. The martensitic transformation in steel represents the most economically significant example of this category of phase transformations but an increasing number of alternatives, such as shape memory alloys, are becoming more important as well.

Classification and Definitions

When a structural change occurs by the coordinated movement of atoms (or groups of atoms) relative to their neighbors then the change is termed *displacive* transformation. This covers a broad range of transformations and so further classifications have been developed [Cohen 1979].

The first distinction can be drawn between transformations dominated by *lattice-distortive strains* and those where *shuffles* are of greater importance.

Homogeneous lattice-distortive strains, also known as Bain strains, are strains that transform one Bravais lattice into a different one. This can be represented by a strain matrix S which transforms one vector, y, into a new vector, x:

$$y = Sx$$

This is homogeneous as straight lines are transformed to new straight lines. Examples of such transformations include a cubic lattice increasing in size on all three axes (dilation) or shearing into a monoclinic structure.

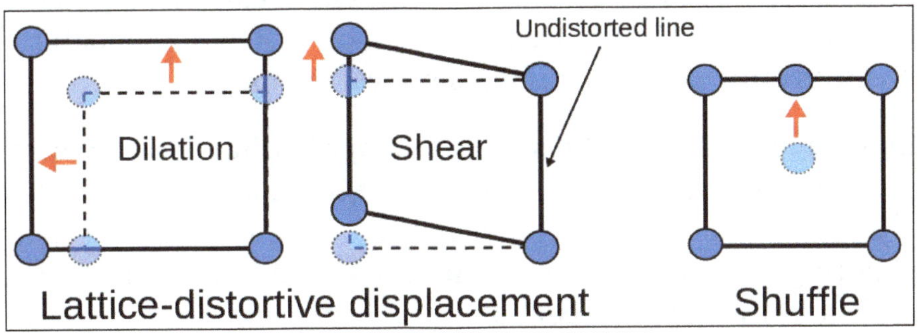

Shuffles, as the name suggests, involve the small movement of atoms within the unit cell. As a result pure shuffles do not normally result in a shape change of the unit cell - only its symmetry and structure.

Phase transformations normally result in the creation of an interface between the transformed and parent material. The energy required to generate this new interface will depend on its nature - essentially how well the two structures fit together. An additional energy term occurs if the transformation includes a shape change since, if new phase is constrained by surrounding material, this may give rise to elastic or plastic deformation and hence a strain energy term. The ratio of these interfacial and strain energy terms has a notable effect on the kinetics of the transformation and the morphology of the new phase. Thus, shuffle transformations, where distortions are small, are dominated by interfacial energies and can be usefully separated from lattice-distortive transformations where the strain energy tends to have a greater effect.

A subclassification of lattice-distortive displacements can be made by considering the dilational and shear components of the distortion. In transformations dominated by the shear component it is possible to find a line in the new phase that is undistorted from the parent phase while all lines are distorted when the dilation is predominant. Shear dominated transformations can be further classified according to the magnitude of the strain energies involved compared to the innate vibrations of the atoms in the lattice and hence whether the strain energies have a notable influence on the kinetics of the transformation and the morphology of the resulting phase. If the strain energy is a significant factor then the transformations are dubbed *martensitic* and if it is not the transformation is referred to as *quasi-martensitic.*

Martensitic Transformation

The difference between austenite and martensite is, in some ways, quite small: while the unit cell of austenite is, on average, a perfect cube, the transformation to martensite distorts this cube by interstitial carbon atoms that do not have time to diffuse out during displacive transformation. The unit cell becomes slightly longer in one dimension and shorter in the other two. The mathematical description of the two structures is quite different, for reasons of symmetry, but the chemical bonding remains very similar. Unlike cementite, which has bonding reminiscent of ceramic materials, the hardness of martensite is difficult to explain in chemical terms.

The explanation hinges on the crystal's subtle change in dimension. Even a microscopic crystallite is millions of unit cells long. Since all of these units face the same direction, distortions of even a fraction of a percent become magnified into a major mismatch between neighboring materials. The mismatch is sorted out by the creation of myriad crystal defects, in a process reminiscent of work hardening. As in work-hardened steel, these defects prevent atoms from sliding past one another in an organized fashion, causing the material to become harder.

Shape memory alloys also have surprising mechanical properties, that were eventually explained by an analogy to martensite. Unlike the iron-carbon system, alloys in the nickel-titanium system can be chosen that make the "martensitic" phase thermodynamically stable.

Pseudomartensitic Transformation

In addition to displacive transformation and diffusive transformation, a new phase transformation that involves displasive sublattice transition and atomic diffusion was discovered using a high-pressure x-ray diffraction system. The new transformation mechanism has been christened a pseudomartensitic transformation.

References

- Lakhani, Archana; Banerjee, A; Chaddah, P; Chen, X; Ramanujan, R V (2012). "Magnetic glass in shape memory alloy: Ni45Co5Mn38Sn12". Journal of Physics: Condensed Matter. 24 (38): 386004. Bibcode:2012JPCM...24L6004L. ISSN 0953-8984. arXiv:1206.2024 . doi:10.1088/0953-8984/24/38/386004
- Modell, Michael; Robert C. Reid (1974). Thermodynamics and Its Applications. Englewood Cliffs, NJ: Prentice-Hall. ISBN 0-13-914861-2
- Krotov, D; Dubuis, J O; Gregor, T; Bialek, W (2014). "Morphogenesis at criticality". PNAS. 111: 3683–3688. doi:10.1073/pnas.1324186111
- Banerjee, A; Pramanik, A K; Kumar, Kranti; Chaddah, P (2006). "Coexisting tunable fractions of glassy and equilibrium long-range-order phases in manganites". Journal of Physics: Condensed Matter. 18 (49): L605. Bibcode:2006JPCM...18L.605B. arXiv:cond-mat/0611152 . doi:10.1088/0953-8984/18/49/L02
- Blundell, Stephen J.; Katherine M. Blundell (2008). Concepts in Thermal Physics. Oxford University Press. ISBN 978-0-19-856770-7
- Mora, Thierry; Bialek, William (2011). "Are biological systems poised at criticality?". Journal of Statistical Physics. 144 (2): 268–302. Bibcode:2011JSP...144..268M. arXiv:1012.2242 . doi:10.1007/s10955-011-0229-4
- Ojovan, Michael I; Lee, William E (2006). "Topologically disordered systems at the glass transition" (PDF). Journal of Physics: Condensed Matter. 18 (50): 11507–11520. Bibcode:2006JPCM...1811507O. doi:10.1088/0953-8984/18/50/007
- Ivancevic, Vladimir G.; Ivancevic, Tijiana, T. (2008). Complex Nonlinearity. Berlin: Springer. pp. 176–177. ISBN 978-3-540-79357-1. Retrieved 19, March 2020
- Jaeger, Gregg (1 May 1998). "The Ehrenfest Classification of Phase Transitions: Introduction and Evolution". Archive for History of Exact Sciences. 53 (1): 51–81. doi:10.1007/s004070050021
- A. D. McNaught and A. Wilkinson (ed.). "Compendium of Chemical Terminology". IUPAC. ISBN 0-86542-684-8. Retrieved 15, August 2020
- YashRoy, R.C. (1987). "13-C NMR studies of lipid fatty acyl chains of chloroplast membranes". Indian Journal of Biochemistry and Biophysics. 24 (6): 177–178
- Leonard, F.; Delamotte, B. (2015). "Critical exponents can be different on the two sides of a transition". Phys. Rev. Lett. 115 (20): 200601. Bibcode:2015PhRvL.115t0601L. arXiv:1508.07852 . doi:10.1103/PhysRevLett.115.200601

6

Metallurgy and its Types

The branch of materials science and engineering that is concerned with the extraction of metals from their ores and modification of metals for use is known as metallurgy. Extractive metallurgy involves the practice of extraction of metals from ores and refining them into a purer form while hydrometallurgy requires aqueous solutions to extract metals from ores. The topics elaborated in this chapter will help in gaining a better perspective about metallurgy.

Metallurgy

Metallurgy is a domain of materials science and engineering that studies the physical and chemical behaviour of metallic elements, their intermetallic compounds, and their mixtures, which are called alloys. Metallurgy is also the technology of metals: the way in which science is applied to the production of metals, and the engineering of metal components for usage in products for consumers and manufacturers. The production of metals involves the processing of ores to extract the metal they contain, and the mixture of metals, sometimes with other elements, to produce alloys. Metallurgy is distinguished from the craft of metalworking, although metalworking relies on metallurgy, as medicine relies on medical science, for technical advancement.

Georgius Agricola, author of *De re metallica*, an important early work on metal extraction.

Metallurgy is subdivided into *ferrous metallurgy* (sometimes also known as black metallurgy) and *non-ferrous metallurgy* or *colored metallurgy*. Ferrous metallurgy involves processes and alloys based on iron while non-ferrous metallurgy involves processes and alloys based on other metals. The production of ferrous metals accounts for 95 percent of world metal production.

Smelting Gold in Nicaragua in the La Luz Gold Mine in Siuna and Bonanza about 1959. Smelting is a basic step in obtaining usable quantities of most metals.

Pouring Smelted Gold into an ingot at the La Luz Gold Mine in Siuna, Nicaragua about 1959.

History

The earliest recorded metal employed by humans appears to be gold, which can be found free or "native". Small amounts of natural gold have been found in Spanish caves used during the late Paleolithic period, c. 40,000 BC. Silver, copper, tin and meteoric iron can also be found in native form, allowing a limited amount of metalworking in early cultures. Egyptian weapons made from meteoric iron in about 3000 BC were highly prized as "daggers from heaven".

Gold headband from Thebes 750–700 BC.

Certain metals, notably tin, lead and (at a higher temperature) copper, can be recovered from their ores by simply heating the rocks in a fire or blast furnace, a process known as smelting. The first evidence of this extractive metallurgy dates from the 5th and 6th millennia BC and was found in the archaeological sites of Majdanpek, Yarmovac and Plocnik, all three in Serbia. To date, the earliest evidence of copper smelting is found at the Belovode site, including a copper axe from 5500 BC belonging to the Vinča culture. Other signs of early metals are found from the third millennium BC in places like Palmela (Portugal), Los Millares (Spain), and Stonehenge (United Kingdom). However, the ultimate beginnings cannot be clearly ascertained and new discoveries are both continuous and ongoing.

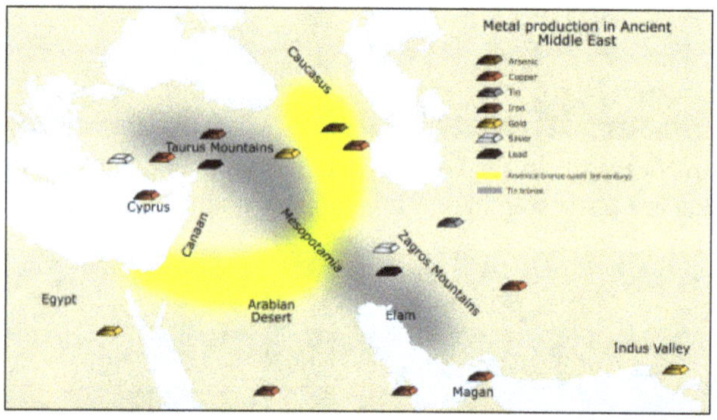

Mining areas of the ancient Middle East. Boxes colors: arsenic is in brown, copper in red, tin in grey, iron in reddish brown, gold in yellow, silver in white and lead in black. Yellow area stands for arsenic bronze, while grey area stands for tin bronze.

These first metals were single ones or as found. About 3500 BC, it was discovered that by combining copper and tin, a superior metal could be made, an alloy called bronze, representing a major technological shift known as the Bronze Age.

The extraction of iron from its ore into a workable metal is much more difficult than for copper or tin. The process appears to have been invented by the Hittites in about 1200 BC, beginning the Iron Age. The secret of extracting and working iron was a key factor in the success of the Philistines.

Historical developments in ferrous metallurgy can be found in a wide variety of past cultures and civilizations. This includes the ancient and medieval kingdoms and empires of the Middle East and Near East, ancient Iran, ancient Egypt, ancient Nubia, and Anatolia (Turkey), Ancient Nok, Carthage, the Greeks and Romans of ancient Europe, medieval Europe, ancient and medieval China, ancient and medieval India, ancient and medieval Japan, amongst others. Many applications, practices, and devices associated or involved in metallurgy were established in ancient China, such as the innovation of the blast furnace, cast iron, hydraulic-powered trip hammers, and double acting piston bellows.

A 16th century book by Georg Agricola called *De re metallica* describes the highly developed and complex processes of mining metal ores, metal extraction and metallurgy of the time. Agricola has been described as the "father of metallurgy".

Extraction

Extractive metallurgy is the practice of removing valuable metals from an ore and refining the extracted raw metals into a purer form. In order to convert a metal oxide or sulfide to a purer metal, the ore must be reduced physically, chemically, or electrolytically.

Furnace bellows operated by waterwheels, Yuan Dynasty, China.

Aluminium plant in Žiar nad Hronom (Central Slovakia).

Extractive metallurgists are interested in three primary streams: feed, concentrate (valuable metal oxide/sulfide), and tailings (waste). After mining, large pieces of the ore feed are broken through crushing and/or grinding in order to obtain particles small enough where each particle is either mostly valuable or mostly waste. Concentrating the particles of value in a form supporting separation enables the desired metal to be removed from waste products.

Mining may not be necessary if the ore body and physical environment are conducive to leaching. Leaching dissolves minerals in an ore body and results in an enriched solution. The solution is collected and processed to extract valuable metals.

Ore bodies often contain more than one valuable metal. Tailings of a previous process may be used as a feed in another process to extract a secondary product from the original ore. Additionally, a concentrate may contain more than one valuable metal. That concentrate would then be processed to separate the valuable metals into individual constituents.

Alloys

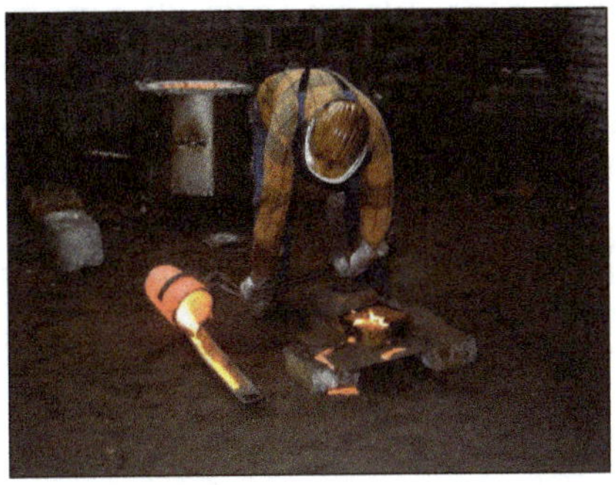

Casting bronze.

Common engineering metals include aluminium, chromium, copper, iron, magnesium, nickel, titanium and zinc. These are most often used as alloys. Much effort has been placed on understanding the iron-carbon alloy system, which includes steels and cast irons. Plain carbon steels (those that contain essentially only carbon as an alloying element) are used in low-cost, high-strength applications where weight and corrosion are not a problem. Cast irons, including ductile iron, are also part of the iron-carbon system.

Stainless steel or galvanized steel are used where resistance to corrosion is important. Aluminium alloys and magnesium alloys are used for applications where strength and lightness are required.

Copper-nickel alloys (such as Monel) are used in highly corrosive environments and for non-magnetic applications. Nickel-based superalloys like Inconel are used in high-temperature applications such as gas turbines, turbochargers, pressure vessels, and heat exchangers. For extremely high temperatures, single crystal alloys are used to minimize creep.

Production

In production engineering, metallurgy is concerned with the production of metallic components for use in consumer or engineering products. This involves the production of alloys, the shaping, the heat treatment and the surface treatment of the product. The task of the metallurgist is to achieve balance between material properties such as cost, weight, strength, toughness, hardness, corrosion, fatigue resistance, and performance in temperature extremes. To achieve this goal, the operating environment must be carefully considered. In a saltwater environment, ferrous metals and some aluminium alloys corrode quickly. Metals exposed to cold or cryogenic conditions may endure a ductile to brittle transition and lose their toughness, becoming more brittle and prone to cracking. Metals under continual cyclic loading can suffer from metal fatigue. Metals under constant stress at elevated temperatures can creep.

Metalworking Processes

Metals are shaped by processes such as:

- Casting – molten metal is poured into a shaped mold.
- Forging – a red-hot billet is hammered into shape.
- Rolling – a billet is passed through successively narrower rollers to create a sheet.
- Laser cladding – metallic powder is blown through a movable laser beam (e.g. mounted on a NC 5-axis machine). The resulting melted metal reaches a substrate to form a melt pool. By moving the laser head, it is possible to stack the tracks and build up a three-dimensional piece.
- Extrusion – a hot and malleable metal is forced under pressure through a die, which shapes it before it cools.
- Sintering – a powdered metal is heated in a non-oxidizing environment after being compressed into a die.

- Machining – lathes, milling machines, and drills cut the cold metal to shape.
- Fabrication – sheets of metal are cut with guillotines or gas cutters and bent and welded into structural shape.
- 3D printing – Sintering or melting powder metal in a very small point on a moving 'print head' moving in 3D space to make any object to shape.

Cold-working processes, in which the product's shape is altered by rolling, fabrication or other processes while the product is cold, can increase the strength of the product by a process called work hardening. Work hardening creates microscopic defects in the metal, which resist further changes of shape.

Various forms of casting exist in industry and academia. These include sand casting, investment casting (also called the "lost wax process"), die casting, and continuous casting.

Heat Treatment

Metals can be heat-treated to alter the properties of strength, ductility, toughness, hardness and/or resistance to corrosion. Common heat treatment processes include annealing, precipitation strengthening, quenching, and tempering. The annealing process softens the metal by heating it and then allowing it to cool very slowly, which gets rid of stresses in the metal and makes the grain structure large and soft-edged so that when the metal is hit or stressed it dents or perhaps bends, rather than breaking; it is also easier to sand, grind, or cut annealed metal. Quenching is the process of cooling a high-carbon steel very quickly after heating, thus "freezing" the steel's molecules in the very hard martensite form, which makes the metal harder. There is a balance between hardness and toughness in any steel; the harder the steel, the less tough or impact-resistant it is, and the more impact-resistant it is, the less hard it is. Tempering relieves stresses in the metal that were caused by the hardening process; tempering makes the metal less hard while making it better able to sustain impacts without breaking.

Often, mechanical and thermal treatments are combined in what are known as thermo-mechanical treatments for better properties and more efficient processing of materials. These processes are common to high-alloy special steels, super alloys and titanium alloys.

Plating

Electroplating is a chemical surface-treatment technique. It involves bonding a thin layer of another metal such as gold, silver, chromium or zinc to the surface of the product. It is used to reduce corrosion as well as to improve the product's aesthetic appearance.

Thermal Spraying

Thermal spraying techniques are another popular finishing option, and often have better high temperature properties than electroplated coatings.

Microstructure

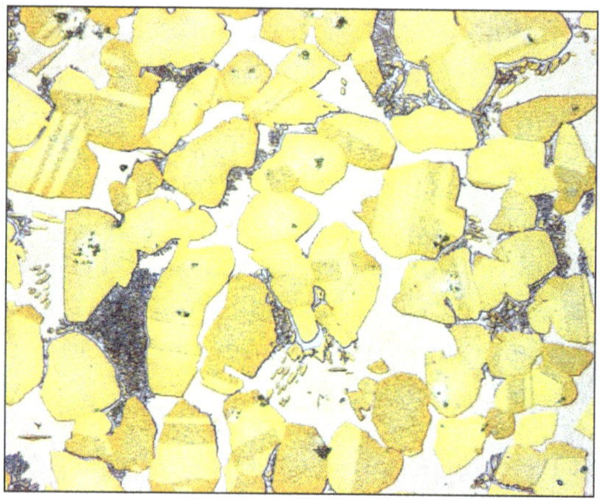

Metallography allows the metallurgist to study the microstructure of metals.

Metallurgists study the microscopic and macroscopic properties using metallography, a technique invented by Henry Clifton Sorby. In metallography, an alloy of interest is ground flat and polished to a mirror finish. The sample can then be etched to reveal the microstructure and macrostructure of the metal. The sample is then examined in an optical or electron microscope, and the image contrast provides details on the composition, mechanical properties, and processing history.

Crystallography, often using diffraction of x-rays or electrons, is another valuable tool available to the modern metallurgist. Crystallography allows identification of unknown materials and reveals the crystal structure of the sample. Quantitative crystallography can be used to calculate the amount of phases present as well as the degree of strain to which a sample has been subjected.

Conferences

EMC, the European Metallurgical Conference has developed to the most important networking business event dedicated to the non-ferrous metals industry in Europe. From the start of the conference sequence in 2001 at Friedrichshafen it was host of the most relevant metallurgists from all countries of the world. The European Metallurgical Conference is organized by GDMB Society of Metallurgists and Miners.

Extractive Metallurgy

Extractive metallurgy is a branch of metallurgical engineering wherein process and methods of extraction of metals from their natural mineral deposits are studied. The field is a materials science, covering all aspects of the types of ore, washing, concentration, separation, chemical processes and extraction of pure metal and their alloying to suit various applications, sometimes for direct use as a finished product, but more often in a form that requires further working to achieve the given properties to suit the applications.

The field of ferrous and non-ferrous extractive metallurgy have specialties that are generically grouped into the categories of mineral processing, hydrometallurgy, pyrometallurgy, and electrometallurgy based on the process adopted to extract the metal. Several processes are used for extraction of same metal depending on occurrence and chemical requirements.

Mineral Processing

Mineral processing begins with beneficiation, consisting of initially breaking down the ore to required sizes depending on the concentration process to be followed, by crushing, grinding, sieving etc. Thereafter, the ore is physically separated from any unwanted impurity, depending on the form of occurrence and/or further process involved. Separation processes take advantage of physical properties of the materials. These physical properties can include density, particle size and shape, electrical and magnetic properties, and surface properties. Major physical and chemical methods include magnetic separation, froth floatation, leaching etc., whereby the impurities and unwanted materials are removed from the ore and the base ore of the metal is concentrated, meaning the percentage of metal in the ore is increased. This concentrate is then either processed to remove moisture or else used as is for extraction of the metal or made into shapes and forms that can undergo further processing, with ease of handling.

Ore bodies often contain more than one valuable metal. Tailings of a previous process may be used as a feed in another process to extract a secondary product from the original ore. Additionally, a concentrate may contain more than one valuable metal. That concentrate would then be processed to separate the valuable metals into individual constituents.

Hydrometallurgy

Hydrometallurgy is concerned with processes involving aqueous solutions to extract metals from ores. The first step in the hydrometallurgical process is leaching, which involves dissolution of the valuable metals into the aqueous solution and /or a suitable solvent. After the solution is separated from the ore solids, the extract is often subjected to various processes of purification and concentration before the valuable metal is recovered either in its metallic state or as a chemical compound. This may include precipitation, distillation, adsorption, and solvent extraction. The final recovery step may involve precipitation, cementation, or an electrometallurgical process. Sometimes, hydrometallurgical processes may be carried out directly on the ore material without any pretreatment steps. More often, the ore must be pretreated by various mineral processing steps, and sometimes by pyrometallurgical processes.

Pyrometallurgy

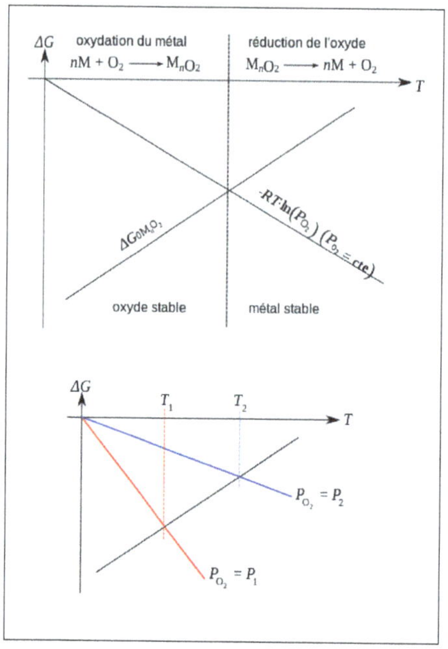

Ellingham diagram for high temperature oxidation.

Pyrometallurgy involves high temperature processes where chemical reactions take place among gases, solids, and molten materials. Solids containing valuable metals are treated to form intermediate compounds for further processing or converted into their elemental or metallic state. Pyrometallurgical processes that involve gases and solids are typified by calcining and roasting operations. Processes that produce molten products are collectively referred to as smelting operations. The energy required to sustain the high temperature pyrometallurgical processes may derive from the exothermic nature of the chemical reactions taking place. Typically, these reactions are oxidation, e.g. of sulfide to sulfur dioxide . Often, however, energy must be added to the process by combustion of fuel or, in the case of some smelting processes, by the direct application of electrical energy.

Ellingham Diagrams are a useful way of analysing the possible reactions, and so predicting their outcome.

Electrometallurgy

Electrometallurgy involves metallurgical processes that take place in some form of electrolytic cell. The most common types of electrometallurgical processes are electrowinning and electro-refining. Electrowinning is an electrolysis process used to recover metals in aqueous solution, usually as the result of an ore having undergone one or more hydrometallurgical processes. The metal of interest is plated onto the cathode, while the anode is an inert electrical conductor. Electro-refining is used to dissolve an impure metallic anode (typically from a smelting process) and produce a high purity cathode. Fused salt electrolysis is another electrometallurgical process whereby the valuable metal has been dissolved into a molten salt which acts as the electrolyte, and the valuable metal collects on the cathode of the cell. The fused salt electrolysis process is conducted at temperatures sufficient to keep both the electrolyte and the metal being produced in the molten state. The scope of electrometallurgy has significant overlap with the areas of hydrometallurgy and (in the case of fused salt electrolysis) pyrometallurgy. Additionally, electrochemical phenomena play a considerable role in many mineral processing and hydrometallurgical processes.

Hydrometallurgy

Hydrometallurgy is a method for obtaining metals from their ores. It is a technique within the field of extractive metallurgy involving the use of aqueous chemistry for the recovery of metals from ores, concentrates, and recycled or residual materials. Metal chemical processing techniques that complement hydrometallurgy are pyrometallurgy, vapour metallurgy and molten salt electrometallurgy. Hydrometallurgy is typically divided into three general areas:

- Leaching.
- Solution concentration and purification.
- Metal or metal compound recovery.

Leaching

Leaching involves the use of aqueous solutions to extract metal from metal bearing materials which is brought into contact with a material containing a valuable metal. The lixiviant solution conditions vary in terms of pH, oxidation-reduction potential, presence of chelating agents and temperature, to optimize the rate, extent and selectivity of dissolution of the desired metal component into the aqueous phase. Through the use of chelating agents, one can selectively extract certain metals. Such chelating agents are typically amines of schiff bases.

The five basic leaching reactor configurations are in-situ, heap, vat, tank and autoclave.

In-situ Leaching

In-situ leaching is also called "solution mining." The process initially involves drilling of holes into the ore deposit. Explosives or hydraulic fracturing are used to create open pathways within the deposit for solution to penetrate into. Leaching solution is pumped into the deposit where it makes contact with the ore. The solution is then collected and processed. The Beverley uranium deposit is an example of in-situ leaching.

Heap Leaching

In heap leaching processes, crushed (and sometimes agglomerated) ore is piled in a heap which is lined with an impervious layer. Leach solution is sprayed over the top of the heap, and allowed to percolate downward through the heap. The heap design usually incorporates collection sumps, which allow the "pregnant" leach solution (i.e. solution with dissolved valuable metals) to be pumped for further processing. An example is gold cyanidation, where pulverized ores are extracted with a solution of sodium cyanide, which, in the presence of air, dissolves the gold, leaving behind the nonprecious residue.

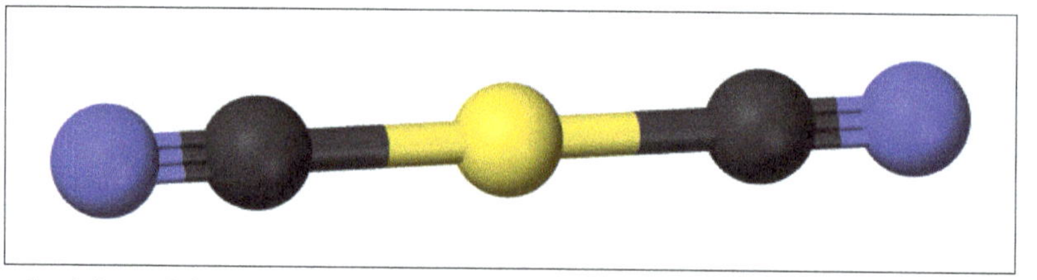

Ball-and-stick model of the aurocyanide or dicyanoaurate(I) complex anion, $[Au(CN)_2]^-$.

Vat Leaching

Vat leaching, also called agitation leaching, involves contacting material, which has usually undergone size reduction and classification, with leach solution in large vats.

Tank Leaching

Stirred tank, also called agitation leaching, involves contacting material, which has usually undergone size reduction and classification, with leach solution in agitated tanks. The agitation can enhance reaction kinetics by enhancing mass transfer. Tanks are often configured as reactors in series.

Autoclave Leaching

Autoclave reactors are used for reactions at higher temperatures, which can enhance the rate of the reaction. Similarly, autoclaved enable the use gaseous reagents in the system.

Solution Concentration and Purification

After leaching, the leach liquor must normally undergo concentration of the metal ions that are to be recovered. Additionally, undesirable metal ions sometimes require removal.

- Precipitation is the selective removal of a compound of the targeted metal or removal of a major impurity by precipitation of one of its compounds. Copper is precipitated as its sulfide as a means to purify nickel leachates.
- Cementation is the conversion of the metal ion to the metal by a redox reaction. A typical application involves addition of scrap iron to a solution of copper ions. Iron dissolves and copper metal is deposited.
- Solvent Extraction.
- Ion Exchange.
- Gas reduction. Treating a solution of nickel and ammonia with hydrogen affords nickel metal as its powder.
- Electrowinning is a particularly selective if expensive electrolysis process applied to the isolation of precious metals. Gold can be electroplated from its solutions.

Solvent Extraction

In the solvent extraction is a mixture of an extractant in a diluent is used to extract a metal from one phase to another. In solvent extraction this mixture is often referred to as the "organic" because the main constituent (diluent) is some type of oil.

The PLS (pregnant leach solution) is mixed to emulsification with the stripped organic and allowed to separate. The metal will be exchanged from the PLS to the organic they are modified. The resulting streams will be a loaded organic and a raffinate. When dealing with electrowinning, the loaded organic is then mixed to emulsification with a lean electrolyte and allowed to separate. The metal will be exchanged from the organic to the electrolyte. The resulting streams will be a stripped organic and a rich electrolyte. The organic stream is recycled through the solvent extraction process while the aqueous streams cycle through leaching and electrowinning processes respectively.

Ion Exchange

Chelating agents, natural zeolite, activated carbon, resins, and liquid organics impregnated with chelating agents are all used to exchange cations or anions with the solution. Selectivity and recovery are a function of the reagents used and the contaminants present.

Metal Recovery

Metal recovery is the final step in a hydrometallurgical process. Metals suitable for sale as raw materials are often directly produced in the metal recovery step. Sometimes, however, further refining is required if ultra-high purity metals are to be produced. The primary types of metal recovery processes are electrolysis, gaseous reduction, and precipitation. For example, a major target of hydrometallurgy is copper, which is conveniently obtained by electrolysis. Cu^{2+} ions reduce at mild potentials, leaving behind other contaminating metals such as Fe^{2+} and Zn^{2+}.

Electrolysis

Electrowinning and electrorefining respectively involve the recovery and purification of metals using electrodeposition of metals at the cathode, and either metal dissolution or a competing oxidation reaction at the anode.

Precipitation

Precipitation in hydrometallurgy involves the chemical precipitation of either metals and their compounds or of the contaminants from aqueous solutions. Precipitation will proceed when, through reagent addition, evaporation, pH change or temperature manipulation, any given species exceeds its limit of solubility.

References

- Seidl, M.; Vuckovic, S.; Gori-Giorgi, P. (2016). "Challenging the Lieb–Oxford bound in a systematic way. Molecular Physics". Molecular Physics. 114 (7-8): 1076–1085. doi:10.1080/00268976.2015.1136440
- Music, D.; Geyer, R.W.; Schneider, J.M. (2016). "Recent progress and new directions in density functional theory based design of hard coatings". Surface & Coatings Technology. 286: 178. doi:10.1016/j.surfcoat.2015.12.021
- Parr, Robert G.; Yang, Weitao (1989). Density-Functional Theory of Atoms and Molecules. New York: Oxford University Press. ISBN 0-19-509276-7

- A.J. Cohen, P. Mori-Sánchez and W. Yang (2012). "Challenges for Density Functional Theory". Chemical Reviews. 112 (1): 289–320. doi:10.1021/cr200107z. PMID 22191548
- Viraht, Xiao-Yin (2012). "Hohenberg-Kohn theorem including electron spin". Physical Review A. 86 (4): 042502. Bibcode:2012PhRvA..86d2502P. doi:10.1103/physreva.86.042502

Permissions

All chapters in this book are published with permission under the Creative Commons Attribution Share Alike License or equivalent. Every chapter published in this book has been scrutinized by our experts. Their significance has been extensively debated. The topics covered herein carry significant information for a comprehensive understanding. They may even be implemented as practical applications or may be referred to as a beginning point for further studies.

We would like to thank the editorial team for lending their expertise to make the book truly unique. They have played a crucial role in the development of this book. Without their invaluable contributions this book wouldn't have been possible. They have made vital efforts to compile up to date information on the varied aspects of this subject to make this book a valuable addition to the collection of many professionals and students.

This book was conceptualized with the vision of imparting up-to-date and integrated information in this field. To ensure the same, a matchless editorial board was set up. Every individual on the board went through rigorous rounds of assessment to prove their worth. After which they invested a large part of their time researching and compiling the most relevant data for our readers.

The editorial board has been involved in producing this book since its inception. They have spent rigorous hours researching and exploring the diverse topics which have resulted in the successful publishing of this book. They have passed on their knowledge of decades through this book. To expedite this challenging task, the publisher supported the team at every step. A small team of assistant editors was also appointed to further simplify the editing procedure and attain best results for the readers.

Apart from the editorial board, the designing team has also invested a significant amount of their time in understanding the subject and creating the most relevant covers. They scrutinized every image to scout for the most suitable representation of the subject and create an appropriate cover for the book.

The publishing team has been an ardent support to the editorial, designing and production team. Their endless efforts to recruit the best for this project, has resulted in the accomplishment of this book. They are a veteran in the field of academics and their pool of knowledge is as vast as their experience in printing. Their expertise and guidance has proved useful at every step. Their uncompromising quality standards have made this book an exceptional effort. Their encouragement from time to time has been an inspiration for everyone.

The publisher and the editorial board hope that this book will prove to be a valuable piece of knowledge for students, practitioners and scholars across the globe.

Index

www.ingramcontent.com/pod-product-compliance
Lightning Source LLC
LaVergne TN
LVHW080730210926
848485LV00003B/542
* 9 7 8 1 6 3 9 8 9 3 4 4 7 *